Pitman Research Notes in Mathematics Series

Submission of proposals for consideration
Suggestions for publication, in the form of outlines and representative samples, are invited by the Editorial Board for assessment. Intending authors should approach one of the main editors or another member of the Editorial Board, citing the relevant AMS subject classifications. Alternatively, outlines may be sent directly to the publisher's offices. Refereeing is by members of the board and other mathematical authorities in the topic concerned, throughout the world.

Preparation of accepted manuscripts
On acceptance of a proposal, the publisher will supply full instructions for the preparation of manuscripts in a form suitable for direct photo-lithographic reproduction. Specially printed grid sheets are provided and a contribution is offered by the publisher towards the cost of typing. Word processor output, subject to the publisher's approval, is also acceptable.

Illustrations should be prepared by the authors, ready for direct reproduction without further improvement. The use of hand-drawn symbols should be avoided wherever possible, in order to maintain maximum clarity of the text.

The publisher will be pleased to give any guidance necessary during the preparation of a typescript, and will be happy to answer any queries.

Important note
In order to avoid later retyping, intending authors are strongly urged not to begin final preparation of a typescript before receiving the publisher's guidelines and special paper. In this way it is hoped to preserve the uniform appearance of the series.

Longman Scientific & Technical
Longman House
Burnt Mill
Harlow, Essex, UK
(tel (0279) 426721)

Titles in this series

1 Improperly posed boundary value problems
A Carasso and A P Stone
2 Lie algebras generated by finite dimensional ideals
I N Stewart
3 Bifurcation problems in nonlinear elasticity
R W Dickey
4 Partial differential equations in the complex domain
D L Colton
5 Quasilinear hyperbolic systems and waves
A Jeffrey
6 Solution of boundary value problems by the method of integral operators
D L Colton
7 Taylor expansions and catastrophes
T Poston and I N Stewart
8 Function theoretic methods in differential equations
R P Gilbert and R J Weinacht
9 Differential topology with a view to applications
D R J Chillingworth
10 Characteristic classes of foliations
H V Pittie
11 Stochastic integration and generalized martingales
A U Kussmaul
12 Zeta-functions: An introduction to algebraic geometry
A D Thomas
13 Explicit *a priori* inequalities with applications to boundary value problems
V G Sigillito
14 Nonlinear diffusion
W E Fitzgibbon III and H F Walker
15 Unsolved problems concerning lattice points
J Hammer
16 Edge-colourings of graphs
S Fiorini and R J Wilson
17 Nonlinear analysis and mechanics: Heriot-Watt Symposium Volume I
R J Knops
18 Actions of fine abelian groups
C Kosniowski
19 Closed graph theorems and webbed spaces
M De Wilde
20 Singular perturbation techniques applied to integro-differential equations
H Grabmüller
21 Retarded functional differential equations: A global point of view
S E A Mohammed
22 Multiparameter spectral theory in Hilbert space
B D Sleeman
24 Mathematical modelling techniques
R Aris
25 Singular points of smooth mappings
C G Gibson
26 Nonlinear evolution equations solvable by the spectral transform
F Calogero
27 Nonlinear analysis and mechanics: Heriot-Watt Symposium Volume II
R J Knops
28 Constructive functional analysis
D S Bridges
29 Elongational flows: Aspects of the behaviour of model elasticoviscous fluids
C J S Petrie
30 Nonlinear analysis and mechanics: Heriot-Watt Symposium Volume III
R J Knops
31 Fractional calculus and integral transforms of generalized functions
A C McBride
32 Complex manifold techniques in theoretical physics
D E Lerner and P D Sommers
33 Hilbert's third problem: scissors congruence
C-H Sah
34 Graph theory and combinatorics
R J Wilson
35 The Tricomi equation with applications to the theory of plane transonic flow
A R Manwell
36 Abstract differential equations
S D Zaidman
37 Advances in twistor theory
L P Hughston and R S Ward
38 Operator theory and functional analysis
I Erdelyi
39 Nonlinear analysis and mechanics: Heriot-Watt Symposium Volume IV
R J Knops
40 Singular systems of differential equations
S L Campbell
41 N-dimensional crystallography
R L E Schwarzenberger
42 Nonlinear partial differential equations in physical problems
D Graffi
43 Shifts and periodicity for right invertible operators
D Przeworska-Rolewicz
44 Rings with chain conditions
A W Chatters and C R Hajarnavis
45 Moduli, deformations and classifications of compact complex manifolds
D Sundararaman
46 Nonlinear problems of analysis in geometry and mechanics
M Atteia, D Bancel and I Gumowski
47 Algorithmic methods in optimal control
W A Gruver and E Sachs
48 Abstract Cauchy problems and functional differential equations
F Kappel and W Schappacher
49 Sequence spaces
W H Ruckle
50 Recent contributions to nonlinear partial differential equations
H Berestycki and H Brezis
51 Subnormal operators
J B Conway

52 Wave propagation in viscoelastic media
F Mainardi
53 Nonlinear partial differential equations and their applications: Collège de France Seminar. Volume I
H Brezis and J L Lions
54 Geometry of Coxeter groups
H Hiller
55 Cusps of Gauss mappings
T Banchoff, T Gaffney and C McCrory
56 An approach to algebraic K-theory
A J Berrick
57 Convex analysis and optimization
J-P Aubin and R B Vintner
58 Convex analysis with applications in the differentiation of convex functions
J R Giles
59 Weak and variational methods for moving boundary problems
C M Elliott and J R Ockendon
60 Nonlinear partial differential equations and their applications: Collège de France Seminar. Volume II
H Brezis and J L Lions
61 Singular systems of differential equations II
S L Campbell
62 Rates of convergence in the central limit theorem
Peter Hall
63 Solution of differential equations by means of one-parameter groups
J M Hill
64 Hankel operators on Hilbert space
S C Power
65 Schrödinger-type operators with continuous spectra
M S P Eastham and H Kalf
66 Recent applications of generalized inverses
S L Campbell
67 Riesz and Fredholm theory in Banach algebra
B A Barnes, G J Murphy, M R F Smyth and T T West
68 Evolution equations and their applications
F Kappel and W Schappacher
69 Generalized solutions of Hamilton-Jacobi equations
P L Lions
70 Nonlinear partial differential equations and their applications: Collège de France Seminar. Volume III
H Brezis and J L Lions
71 Spectral theory and wave operators for the Schrödinger equation
A M Berthier
72 Approximation of Hilbert space operators I
D A Herrero
73 Vector valued Nevanlinna Theory
H J W Ziegler
74 Instability, nonexistence and weighted energy methods in fluid dynamics and related theories
B Straughan
75 Local bifurcation and symmetry
A Vanderbauwhede
76 Clifford analysis
F Brackx, R Delanghe and F Sommen
77 Nonlinear equivalence, reduction of PDEs to ODEs and fast convergent numerical methods
E E Rosinger
78 Free boundary problems, theory and applications. Volume I
A Fasano and M Primicerio
79 Free boundary problems, theory and applications. Volume II
A Fasano and M Primicerio
80 Symplectic geometry
A Crumeyrolle and J Grifone
81 An algorithmic analysis of a communication model with retransmission of flawed messages
D M Lucantoni
82 Geometric games and their applications
W H Ruckle
83 Additive groups of rings
S Feigelstock
84 Nonlinear partial differential equations and their applications: Collège de France Seminar. Volume IV
H Brezis and J L Lions
85 Multiplicative functionals on topological algebras
T Husain
86 Hamilton-Jacobi equations in Hilbert spaces
V Barbu and G Da Prato
87 Harmonic maps with symmetry, harmonic morphisms and deformations of metrics
P Baird
88 Similarity solutions of nonlinear partial differential equations
L Dresner
89 Contributions to nonlinear partial differential equations
C Bardos, A Damlamian, J I Díaz and J Hernández
90 Banach and Hilbert spaces of vector-valued functions
J Burbea and P Masani
91 Control and observation of neutral systems
D Salamon
92 Banach bundles, Banach modules and automorphisms of C*-algebras
M J Dupré and R M Gillette
93 Nonlinear partial differential equations and their applications: Collège de France Seminar. Volume V
H Brezis and J L Lions
94 Computer algebra in applied mathematics: an introduction to MACSYMA
R H Rand
95 Advances in nonlinear waves. Volume I
L Debnath
96 FC-groups
M J Tomkinson
97 Topics in relaxation and ellipsoidal methods
M Akgül
98 Analogue of the group algebra for topological semigroups
H Dzinotyiweyi
99 Stochastic functional differential equations
S E A Mohammed

100 Optimal control of variational inequalities
V Barbu
101 Partial differential equations and dynamical systems
W E Fitzgibbon III
102 Approximation of Hilbert space operators. Volume II
C Apostol, L A Fialkow, D A Herrero and D Voiculescu
103 Nondiscrete induction and iterative processes
V Ptak and F-A Potra
104 Analytic functions – growth aspects
O P Juneja and G P Kapoor
105 Theory of Tikhonov regularization for Fredholm equations of the first kind
C W Groetsch
106 Nonlinear partial differential equations and free boundaries. Volume I
J I Díaz
107 Tight and taut immersions of manifolds
T E Cecil and P J Ryan
108 A layering method for viscous, incompressible L_p flows occupying R^n
A Douglis and E B Fabes
109 Nonlinear partial differential equations and their applications: Collège de France Seminar. Volume VI
H Brezis and J L Lions
110 Finite generalized quadrangles
S E Payne and J A Thas
111 Advances in nonlinear waves. Volume II
L Debnath
112 Topics in several complex variables
E Ramírez de Arellano and D Sundararaman
113 Differential equations, flow invariance and applications
N H Pavel
114 Geometrical combinatorics
F C Holroyd and R J Wilson
115 Generators of strongly continuous semigroups
J A van Casteren
116 Growth of algebras and Gelfand–Kirillov dimension
G R Krause and T H Lenagan
117 Theory of bases and cones
P K Kamthan and M Gupta
118 Linear groups and permutations
A R Camina and E A Whelan
119 General Wiener–Hopf factorization methods
F-O Speck
120 Free boundary problems: applications and theory, Volume III
A Bossavit, A Damlamian and M Fremond
121 Free boundary problems: applications and theory, Volume IV
A Bossavit, A Damlamian and M Fremond
122 Nonlinear partial differential equations and their applications: Collège de France Seminar. Volume VII
H Brezis and J L Lions
123 Geometric methods in operator algebras
H Araki and E G Effros
124 Infinite dimensional analysis–stochastic processes
S Albeverio
125 Ennio de Giorgi Colloquium
P Krée
126 Almost-periodic functions in abstract spaces
S Zaidman
127 Nonlinear variational problems
A Marino, L Modica, S Spagnolo and M Degiovanni
128 Second-order systems of partial differential equations in the plane
L K Hua, W Lin and C-Q Wu
129 Asymptotics of high-order ordinary differential equations
R B Paris and A D Wood
130 Stochastic differential equations
R Wu
131 Differential geometry
L A Cordero
132 Nonlinear differential equations
J K Hale and P Martinez-Amores
133 Approximation theory and applications
S P Singh
134 Near-rings and their links with groups
J D P Meldrum
135 Estimating eigenvalues with *a posteriori/a priori* inequalities
J R Kuttler and V G Sigillito
136 Regular semigroups as extensions
F J Pastijn and M Petrich
137 Representations of rank one Lie groups
D H Collingwood
138 Fractional calculus
G F Roach and A C McBride
139 Hamilton's principle in continuum mechanics
A Bedford
140 Numerical analysis
D F Griffiths and G A Watson
141 Semigroups, theory and applications. Volume I
H Brezis, M G Crandall and F Kappel
142 Distribution theorems of L-functions
D Joyner
143 Recent developments in structured continua
D De Kee and P Kaloni
144 Functional analysis and two-point differential operators
J Locker
145 Numerical methods for partial differential equations
S I Hariharan and T H Moulden
146 Completely bounded maps and dilations
V I Paulsen
147 Harmonic analysis on the Heisenberg nilpotent Lie group
W Schempp
148 Contributions to modern calculus of variations
L Cesari
149 Nonlinear parabolic equations: qualitative properties of solutions
L Boccardo and A Tesei
150 From local times to global geometry, control and physics
K D Elworthy

151 A stochastic maximum principle for optimal control of diffusions
U G Haussmann
152 Semigroups, theory and applications. Volume II
H Brezis, M G Crandall and F Kappel
153 A general theory of integration in function spaces
P Muldowney
154 Oakland Conference on partial differential equations and applied mathematics
L R Bragg and J W Dettman
155 Contributions to nonlinear partial differential equations. Volume II
J I Díaz and P L Lions
156 Semigroups of linear operators: an introduction
A C McBride
157 Ordinary and partial differential equations
B D Sleeman and R J Jarvis
158 Hyperbolic equations
F Colombini and M K V Murthy
159 Linear topologies on a ring: an overview
J S Golan
160 Dynamical systems and bifurcation theory
M I Camacho, M J Pacifico and F Takens
161 Branched coverings and algebraic functions
M Namba
162 Perturbation bounds for matrix eigenvalues
R Bhatia
163 Defect minimization in operator equations: theory and applications
R Reemtsen
164 Multidimensional Brownian excursions and potential theory
K Burdzy
165 Viscosity solutions and optimal control
R J Elliott
166 Nonlinear partial differential equations and their applications. Collège de France Seminar. Volume VIII
H Brezis and J L Lions
167 Theory and applications of inverse problems
H Haario
168 Energy stability and convection
G P Galdi and B Straughan
169 Additive groups of rings. Volume II
S Feigelstock
170 Numerical analysis 1987
D F Griffiths and G A Watson
171 Surveys of some recent results in operator theory. Volume I
J B Conway and B B Morrel
172 Amenable Banach algebras
J-P Pier
173 Pseudo-orbits of contact forms
A Bahri
174 Poisson algebras and Poisson manifolds
K H Bhaskara and K Viswanath
175 Maximum principles and eigenvalue problems in partial differential equations
P W Schaefer
176 Mathematical analysis of nonlinear, dynamic processes
K U Grusa
177 Cordes' two-parameter spectral representation theory
D F McGhee and R H Picard
178 Equivariant K-theory for proper actions
N C Phillips
179 Elliptic operators, topology and asymptotic methods
J Roe
180 Nonlinear evolution equations
J K Engelbrecht, V E Fridman and E N Pelinovski
181 Nonlinear partial differential equations and their applications. Collège de France Seminar. Volume IX
H Brezis and J L Lions
182 Critical points at infinity in some variational problems
A Bahri
183 Recent developments in hyperbolic equations
L Cattabriga, F Colombini, M K V Murthy and S Spagnolo
184 Optimization and identification of systems governed by evolution equations on Banach space
N U Ahmed
185 Free boundary problems: theory and applications. Volume I
K H Hoffmann and J Sprekels
186 Free boundary problems: theory and applications. Volume II
K H Hoffmann and J Sprekels
187 An introduction to intersection homology theory
F Kirwan
188 Derivatives, nuclei and dimensions on the frame of torsion theories
J S Golan and H Simmons
189 Theory of reproducing kernels and its applications
S Saitoh
190 Volterra integrodifferential equations in Banach spaces and applications
G Da Prato and M Iannelli
191 Nest algebras
K R Davidson
192 Surveys of some recent results in operator theory. Volume II
J B Conway and B B Morrel
193 Nonlinear variational problems. Volume II
A Marino and M K Murthy
194 Stochastic processes with multidimensional parameter
M E Dozzi
195 Prestressed bodies
D Iesan
196 Hilbert space approach to some classical transforms
R H Picard
197 Stochastic calculus in application
J R Norris
198 Radical theory
B J Gardner
199 The C* – algebras of a class of solvable Lie groups
X Wang

200 Stochastic analysis, path integration and dynamics
D Elworthy
201 Riemannian geometry and holonomy groups
S Salamon
202 Strong asymptotics for extremal errors and polynomials associated with Erdös type weights
D S Lubinsky
203 Optimal control of diffusion processes
V S Borkar
204 Rings, modules and radicals
B J Gardner
205 Numerical studies for nonlinear Schrödinger equations
B M Herbst and J A C Weideman
206 Distributions and analytic functions
R D Carmichael and D Mitrović
207 Semicontinuity, relaxation and integral representation in the calculus of variations
G Buttazzo
208 Recent advances in nonlinear elliptic and parabolic problems
P Bénilan, M Chipot, L Evans and M Pierre
209 Model completions, ring representations and the topology of the Pierce sheaf
A Carson
210 Retarded dynamical systems
G Stepan
211 Function spaces, differential operators and nonlinear analysis
L Paivarinta
212 Analytic function theory of one complex variable
C C Yang, Y Komatu and K Niino
213 Elements of stability of visco-elastic fluids
J Dunwoody
214 Jordan decompositions of generalised vector measures
K D Schmidt
215 A mathematical analysis of bending of plates with transverse shear deformation
C Constanda
216 Ordinary and partial differential equations Vol II
B D Sleeman and R J Jarvis
217 Hilbert modules over function algebras
R G Douglas and V I Paulsen
218 Graph colourings
R Wilson and R Nelson
219 Hardy-type inequalities
A Kufner and B Opic
220 Nonlinear partial differential equations and their applications. College de France Seminar Volume X
H Brezis and J L Lions
221 Workshop on dynamical systems
E Shiels and Z Coelho
222 Geometry and analysis in nonlinear dynamics
H W Broer and F Takens
223 Fluid dynamical aspects of combustion theory
M Onofri and A Tesei
224 Approximation of Hilbert space operators. Volume I. 2nd edition
D Herrero
225 Operator Theory: Proceedings of the 1988 GPOTS–Wabash conference
J B Conway and B B Morrel
226 Local cohomology and localization
J L Bueso Montero, B Torrecillas Jover and A Verschoren
227 Sobolev spaces of holomorphic functions
F Beatrous and J Burbea
228 Numerical analysis. Volume III
D F Griffiths and G A Watson
229 Recent developments in structured continua. Volume III
D De Kee and P Kaloni
230 Boolean methods in interpolation and approximation
F J Delvos and W Schempp
231 Further advances in twistor theory, Volume 1
L J Mason and L P Hughston
232 Further advances in twistor theory, Volume 2
L J Mason and L P Hughston
233 Geometry in the neighborhood of invariant manifolds of maps and flows and linearization
U Kirchgraber and K Palmer
234 Quantales and their applications
K I Rosenthal
235 Integral equations and inverse problems
R Lazarov and V Petkov
236 Pseudo-differential operators
S R Simanca
237 A functional analytic approach to statistical experiments
I M Bomze
238 Quantum mechanics, algebras and distributions
D Dubin and M Hennings
239 Hamilton flows and evolution semigroups
J Gzyl

V S Borkar

Indian Institute of Science, Bangalore

Topics in Controlled Markov Chains

Copublished in the United States with
John Wiley & Sons, Inc., New York

Longman Scientific & Technical,
Longman Group UK Limited,
Longman House, Burnt Mill, Harlow,
Essex CM20 2JE, England
and Associated Companies throughout the world.

Copublished in the United States with
John Wiley & Sons, Inc., 605 Third Avenue, New York, NY 10158

First published 1991

AMS Subject Classification: 93E20, 60J10

ISSN 0269-3674

British Library Cataloguing in Publication Data
Borkar, Vivek S.
Topics in controlled Markov chains.
1. Markov Chains
I. Title
519.233

ISBN 0-582-06821-5

Library of Congress Cataloging-in-Publication Data
Borkar, Vivek S.
Topics in controlled Markov chains / Vivek S. Borkar.
p. cm.—(Pitman research notes in mathematics series, ISSN
ISSN 0269-3674; 240)
Includes bibliographical references.
1. Markov chains. I. Title. II. Series.
QA274.7.B67 1990
519.2′33--dc20 90-42154
CIP

Printed and bound in Great Britain
by Biddles Ltd, Guildford and King's Lynn

For my parents

Shripad and Sarita Borkar

Contents

Preface

I was introduced to controlled Markov chains as a graduate student of electrical engineering at the University of California, Berkeley. Ever since, it has been my favourite topic, despite working for the most part in an environment where such work is not really 'kosher'. This book is essentially a summing up of my contributions to the field over the past decade. As such, it is a somewhat idiosyncratic account of some old results cast in new moulds along with some new results. I hope it will have its role to play in the future development of the subject.

The work reported here was done in several places, and I would like to thank the institutions for their support: Department of Electrical Engineering and Computer Sciences of the University of California, Berkeley, where some of the early work on adaptive control reported here was done; Tata Institute of Fundamental Research, Bangalore Center, where the bulk of the work was done; and the following places, where I completed parts of this work during my visits: Technische Hogeschool Twente, Holland; The Institute for Mathematics and its Applications at the University of Minnesota, Minneapolis; and the Systems Research Center, University of Maryland, College Park. A part of the work was presented in a series of seminars at the last mentioned institution.

I would like to thank Mrinal Ghosh and P. Guruswamy Babu for carefully reading the manuscript. Thanks are also due to Ms. K.S. Vishalakshi and Mrs. Shantha Srinivas for typing the first draft and Ms. Terri Moss for preparing the final version.

Bangalore, July, 1989. Vivek S. Borkar

Introduction

"....... you have nothing to lose but your chains".

The study of controlled Markov chains on a countable state space is an old and mature field by modern standards. However, it is far from exhausted, as the extent of ongoing activity indicates. In recent times, much of this has been spurred by new applications to queuing systems, computer communications etc. which often stretch the limits of the existing theory and force new developments. These trends are well reflected in the recent texts, [KuVa], [Whi 1], [Whi 2], [Tijm] among others.

The aim of the present monograph is more modest. It gives a unified account (and with hindsight, a considerably streamlined one) of author's own contributions to the subject over the past decade [Bor 1]-[BoV 3]. This splits into two parts. The first may be loosely termed 're-laying the foundations' and the second 'special topics'.

The former represents an attempt to provide an alternative framework for the study of classical problems in controlled Markov chains based on convex analysis. The key feature of this approach is the formulation of the control problem as an optimization problem on a suitably defined convex set of 'occupation measures'. We elaborate on this point below.

Consider for example the infinite horizon discounted cost control problem. The cost for this problem can be viewed as the integral of the running cost function defined on the product of state and control spaces with respect to a naturally defined discounted occupation measure on this space. The attainable set of these measures over all control strategies can be shown to be compact convex, with its extreme points corresponding to stationary strategies. The existence of an optimal stationary strategy is then assured without the help of the conventional dynamic programming heuristic. The dynamic programming equations in turn can be recovered from the foregoing by a small amount of additional work.

As a second example, consider the 'ergodic' or 'long-run average cost' control problem. For the sake of simplicity, assume that the running cost depends only on the current state and does not depend explicitly on the control. The cost then is the limit (more generally, limsup) of the integral of the running cost with respect to the empirical measures of the chain. Consider the latter as probability measures on the one point compactification of the state space. One can then show that with probability one, the restriction of any limit point of these measures to the original state space is a scalar multiple of the invariant probability measure under some stable stationary randomized strategy. The set of such invariant probability measures is in turn a closed convex set whose extreme points correspond to stable stationary strategies. This gives us the existence of an optimal stable stationary strategy under suitable additional hypotheses. The optimality here is in the strong (i.e. almost sure) sense. This can be made a starting point of a study of the associated dynamic programming equations by a direct method that avoids the traditional 'vanishing discount' argument.

In the first instance above, what we achieve is an elegant reformulation of the problem that offers a lot of additional insight. In the second case the gain is much more substantial: we are able to handle the ergodic control problem in a vastly more general set-up than has been possible so far.

However, this extra ground gained in the classical terrain is only a secondary motivation for proposing the new convex analytic approach. The primary impetus comes from the possibilities it opens up in the nonclassical domains such as multiobjective control, where the traditional dynamic programming heuristic has obvious limitations. Our one success story in this direction is what constitutes the first of the 'special topics'. This is a constrained control problem introduced by Beutler and Ross [BeRo], [Ross] which we recast as a convex analysis problem and develop a suitable existence theory for. Some related multiobjective control problems are also discussed *en passant*.

The second special topic considered is the problem of control under partial observations. We introduce a new state variable for this problem, called the unnormalized conditional law, which offers some computational advantages. We mostly mimic the corresponding developments for the complete observations case.

The third and final special topic considered is a problem in adaptive control which will be treated in considerable detail. This is the self-tuning scheme of Mandl [Mand] for the parametric, non-Bayesian adaptive control of Markov chains. This scheme employs an on-line parameter estimation scheme such as the maximum likelihood for estimating the unknown parameter, and picks at each time the control that would have been the optimal choice were the current parameter estimate the true parameter. Almost sure asymptotic optimality of this scheme for the ergodic control problem is proved under a suitable 'identifiability condition'. The latter condition is restrictive in practice. A clever way to deal with the problems caused by its absence is an explicit cost bias in the estimation scheme suggested by Kumar, Becker and Lin [KuBe], [KuLi]. This variant of the self-tuner and some related heuristic issues are also discussed.

The material is organized as follows:

Chapter I surveys the basic theory of Markov chains on a countable state space. This is standard material. My rendition of it is largely taken from my notes of some informal lectures given by Jean Walrand during our joint student days at Berkeley.

Chapter II introduces the terminology and notation of [Bor 1]-[Bo Gh], which is followed throughout the book. This notation is not standard, it takes some getting used to. The effort is well worth it because of the economy and clarity that it offers later on. The chapter also introduces the classical control problems and concludes with some general results concerning controlled Markov chains.

Our study of the classical control problems begins with Chapters III and IV, which consider the three 'cumulative cost' problems: the infinite horizon discounted cost problem (Chapter III), the problem of control up to an exit time, and the finite horizon control problem (Chapter IV). As already indicated, these are studied from an alternative, convex analytic point of view.

The ergodic control problem is structurally different from the above problems as it involves only the asymptotic time-averaged behaviour of the process, its finite time behaviour being completely irrelevant. It thus requires a separate and more involved treatment. We devote two chapters to this problem. Chapter V derives the basic existence results for optimal stable stationary strategies based on a characterization of almost sure limit points for the empirical measures of the joint state and control process. Chapter VI studies the associated dynamic programming equations.

The bulk of Chapter VII is devoted to the Beutler-Ross problem mentioned above. The rest discusses some general issues in multiobjective control.

Chapter VIII gives a novel treatment of the control problems under partial observations, based on a new state variable and the convex analytic approach. Only the essentials are sketched since the details largely imitate the corresponding developments for the complete observations case.

Chapters IX and X are devoted to some specific problems in parametric, non-Bayesian adaptive control of Markov chains. The first of these chapters begins with a study of the maximum likelihood estimates for the unknown parameter entering the controlled Markov chain, and uses these results to prove the almost sure optimality of Mandl's self-tuning scheme described above for the ergodic control problem. This, however, is possible only under a strong 'identifiability condition' which requires that any two distinct models be discriminated from each other by any control strategy. In the absence of such a condition, only weaker results are possible even in the relatively simple situation of the possible set of parameters being finite. The Kumar-Becker-Lin device of biasing the log-likelihood function to favour parameters leading to lower optimal cost avoids this problem. This is described at length in Chapter X.

As already announced at the beginning of this introduction, this book has the limited aim of presenting the author's work in this field. Needless to say, it leaves out many interesting and important recent developments. As a compensation (admittedly inadequate), a brief list of some concurrent developments of interest is given in Chapter XI with representative references. This chapter also collects together some of the open issues raised above.

The theory of spaces of probability measures on Polish spaces with the Prohorov topology plays a key role in Chapters II-X above. As this does not always form a part of an applied probabilist's training (at least to the level of sophistication required here), a brief account of this is presented in the Appendix.

Each chapter is preceded by a brief preamble describing its contents. The sections in, say, Chapter X are numbered X.1, X.2, ... and so on. The theorems, equations etc. in section X.m will be numbered as Theorem m.n, (m.n) and so on. They will be referred to by this number within the chapter, and by Theorem X.m.n., (X.m.n) in other chapters.

Chapter I
Markov Chains: A Review

This chapter briefly surveys the fundamentals of Markov chains on a countable state space, especially the strong Markov property, the classification of states, and the ergodic behaviour.

I.1 Construction and the Strong Markov Property

Let S be a countable set, labelled $\{1,2,...\}$. Let $\lambda \in \mathbb{P}(S)$. (Here and in the rest of this book, $\mathbb{P}(X)$ for a Polish space X will denote the Polish space of probability measures on X with the Prohorov topology. See appendix.) Let $P = [\,[p(i, j)]\,]$, $i, j \in S$, be a possibly infinite matrix satisfying

$$p(i, j) \in [0,1], \sum_{k \in S} p(i, k) = 1, \quad i, j \in S. \tag{1.1}$$

A sequence $\{X_n, n = 0,1,2,...\}$ of S-valued random variables is called a Markov chain with state space S, initial distribution λ and transition matrix P if

(a) $P(X_0 \in A) = \lambda(A)$ $\qquad \forall\ A \subset S,\ n \geq 0,$

(b) $P(X_{n+1} = j/X_n = i) = p(i, j)$ $\qquad \forall\ i, j \in S,\ n \geq 0,$

(c) $P(X_{n+1} \in A/F_n) = P(X_{n+1} \in A/X_n)$ $\qquad \forall\ A \subset S,\ n \geq 0,$

where $F_n = \sigma(X_0, X_1,...,X_n)$.

Strictly speaking $\{X_n\}$ above defines only a time-homogeneous Markov chain or a Markov chain with stationary transitions. We shall comment on this qualification later. Property (c) above is called the Markov property and easily generalizes to

(c_1) $P((X_n, X_{n+1}, X_{n+}),..., \in B/F_n) = P((X_n, X_{n+1}, X_{n+2},...) \in B/X_n)\ \forall\ n \geq 0$
and $B \in$ the product σ-field of $S^\infty = S \times S \times S \times ...$

On identifying the subscript n above with discrete time steps, (c_1) becomes equivalent to the following statement: At time n, the 'future' $(X_{n+1}, X_{n+2},...)$ is conditionally independent of the 'past' $(X_0, X_1,...,X_{n-1})$ given the 'present' i.e., X_n. Since this statement is symmetric in time, (c_1) is also equivalent to

(c_2) $P((X_0, X_1,...,X_n) \in B/X_n, X_{n+1},...) = P((X_0, X_1,...,X_n) \in B/X_n)\ \forall\ n \geq 0,$
$B \subset S^{n+1} = S \times S \times ... \times S$ (n + 1 times).

Given S, λ, P as above, one can construct a Markov chain $\{X_n\}$ satisfying (a)-(c) above on the 'canonical' space $\Omega = S^\infty$ endowed with the product σ-field F. One proceeds as follows: Let $A \in F$ be of the form $A = \prod_{i=0}^{\infty} A_i$ where $A_i \subset S$ and $A_i \neq S$ for at most finitely many i. Write $\lambda(j) = \lambda(\{j\})$ for $j \in S$.

Define a probability measure P on (Ω, F) by

$$P(A) = \sum_{j_0 \in A_0} \sum_{j_1 \in A_1} \cdots \sum_{j_n \in A_n} \lambda(j_0)p(j_0, j_1)p(j_1, j_2)p(j_{n-1}, j_n)$$

for any $n \geq 1$ for which $i > n$ implies $A_i = S$. Using (1.1), one can easily check that this definition is independent of the specific choice of n. This in fact defines a probability measure only on the Boolean algebra formed by sets A of the type above (the 'cylinder sets'). By Kolmogorov's extension theorem, it extends uniquely to a probability measure P on (Ω, F). For $n \geq 0$ and $\omega = (\omega_0, \omega_1, \omega_2,...) \in \Omega$, define $X_n(\omega) = \omega_n$. Then $\{X_n\}$ is the desired Markov chain. From here on, we shall assume this canonical set-up.

In particular, it is clear that the law of $\{X_n\}$ is uniquely specified by λ and P. Let $P_j \in \mathbb{P}(\Omega)$ denote the law of $(X_0, X_1, X_2, ...)$ when λ is the Dirac measure at $j \in S$. Then (c_1) can be rewritten as follows:

(c_3) $P((X_n, X_{n+1}, ...) \in B/F_n) = P_{X_n}((X_0, X_1, ...) \in B)\ \ \forall\ n \geq 0,\ B \in F.$

$\{X_n\}$ is said to have the strong Markov property (SMP for short) if for any $\{F_n\}$-stopping time τ,

$$I\{\tau < \infty\}P((X_\tau, X_{\tau+1},...) \in B/F_\tau) = I\{\tau < \infty\}P_{X_\tau}((X_0, X_1,...) \in B) \text{ a.s.}$$

for all $B \in F$. Write $\varphi(j) = P_j((X_0, X_1,...) \in B)$ for $j \in S$ and some prescribed $B \in F$.

Theorem 1.1. $\{X_n\}$ has the SMP.

Proof. It suffices to show that for all $A \in F_\tau$,

$$\int_A I\{\tau < \infty\}P((X_\tau, X_{\tau+1},...) \in B/F_\tau)dP =$$

$$\int_A I\{\tau < \infty\}P_{X_\tau}((X_0, X_1,...) \in B)dP. \tag{1.2}$$

The left-hand side of (1.2) equals

$$P(\{(X_\tau, X_{\tau+1}, ...) \in B\} \cap A \cap \{\tau < \infty\})$$

$$= \sum_n P(\{(X_n, X_{n+1}, ...) \in B\} \cap A \cap \{\tau = n\})$$

$$= \sum_n E[P((X_n, X_{n+1}, ...) \in B/F_n) I_{A \cap \{\tau = n\}}]$$

$$= \sum_n \int_{A \cap \{\tau = n\}} \varphi(X_n) dP$$

..... by (c_1)

$$= \int_{A \cap \{\tau < \infty\}} \varphi(X_\tau) dP,$$

which equals the right-hand side of (1.2). QED.

The statement of the SMP can be further simplified as follows: Let $\theta_n : S^\infty \to S^\infty$ denote the measurable map $(w_0, w_1, w_2, ...) \to (w_n, w_{n+1}, ...)$, $n \geq 0$. Then $X_m \circ \theta_n = X_{m+n}$ in the canonical set-up. For an $\{F_n\}$-stopping time τ, write $\theta_\tau(w) = \theta_{\tau(w)}(w)$ for $w \in \{\tau < \infty\}$. For $A \in F_\tau$ satisfying $A \subset \{X_\tau = j\}$ for some $j \in S$, SMP implies that for any $B \in F$,

$$P(\{\theta_\tau(w) \in B\} \cap A) = \int_A P(\theta_\tau(w) \in B/X_\tau) dP = P_j(B)P(A).$$

That is,

$$P(A \cap \theta_\tau^{-1}(B)) = P(A)P_j(B). \tag{1.3}$$

Conversely, (1.3) implies the SMP, as we prove next. Let $A \in F_\tau$, $B \in F$, and set $A_i = A \cap \{X_\tau = i\}$, $i \geq 0$. Then

$$P(\{(X_\tau, X_{\tau+1}, ...) \in B\} \cap A_i) = P_i(B)P(A_i)$$

by (1.3) and hence

$$P(\{(X_\tau, X_{\tau+1}, ...) \in B\} \cap A) = \sum_i P_i(B)P(A_i)$$

$$= \sum_i \int_{A_i} P_i(B)dP$$

$$= \sum_i \int_{A_i} P_{X_\tau}(B)dP$$

$$= \int_A P_{X_\tau}(B)dP,$$

implying the SMP. Thus (1.3) is an alternative statement of the SMP.

Let $Y \in L_1(\Omega, F, P)$. One important consequence of the SMP is that

$$E[Y \circ \theta_\tau / F_\tau] = E_{X_\tau}[Y]$$

where $E_j[\cdot]$ denotes the expectation under P_j. To see this, first note that it holds for indicator functions, hence for simple functions and by the conditional monotone convergence theorem, for non-negative integrable functions. The claim for a general integrable random variable then follows by the standard gimmick of writing it as a difference of two non-negative integrable random variables.

II.2 Classification of States

In this section, we classify the elements of S according to certain recurrence properties to be defined.

Let $\tau_j = \min\{n > 0 \mid X_n = j\}$ for $j \in S$ denote the 'first hitting time' of j ('first return time' if $X_0 = j$). Define $\rho_{ij} = P_i(\tau_j < \infty)$, $N_j = \sum_{n=1}^{\infty} I\{X_n = j\}$ and $G(i, j) = E_i[N_j]$. (N_j and G(i, j) can be $+\infty$.) Letting P^n denote the n times matrix product of P with itself and $p^n(i, j)$ its (i,j)th element, we have

$$G(i, j) = \sum_{n=1}^{\infty} p^n(i,j).$$

Lemma 2.1. $$P_i(N_j = m) = \rho_{ij}\, \rho_{jj}^{m-1}(1 - \rho_{jj}),\ m \geq 1.$$

Proof. Let $\sigma_1 = \tau_j$, $\sigma_n = \min\{m > \sigma_{n-1} \mid X_m = j\}$ for n = 2,3,... . Then

$$P_i(N_j = m) = P_i(\sigma_\ell < \infty,\ 1 \le \ell \le m,\ \sigma_{m+1} = \infty)$$

$$= P_i(\tau_j < \infty)\underbrace{P_j(\tau_j < \infty) \ldots P_j(\tau_j < \infty)}_{(m-1)\text{times}} P_j(\tau_j = \infty)$$

$$= \rho_{ij}\, \rho_{jj}^{m-1} (1 - \rho_{jj})$$

where the second step follows by repeated application of the SMP. QED.

Theorem 2.1. If $\rho_{jj} < 1$, $P_i(N_j = \infty) = 0$ and $G(i, j) < \infty$. If $\rho_{jj} = 1$, $P_j(N_j = \infty) = 1$, $G(j, j) = \infty$, $P_i(N_j = \infty) = \rho_{ij} = 1 - P_i(N_j = 0)$ and $G(i, j) = \infty$ or 0 according to whether $\rho_{ij} > 0$ or $= 0$.

Proof. This is immediate from the preceding Lemma. QED.

Let $S_T = \{j \in S \mid \rho_{jj} < 1\}$ and $S_R = \{j \in S \mid \rho_{jj} = 1\}$. The elements of S_T and S_R are called transient and recurrent states respectively. Write $i \to j$ whenever $\rho_{ij} > 0$.

Theorem 2.2. If $i \in S_R$ and $i \to j$, then $j \in S_R$ and $\rho_{ij} = \rho_{ji} = 1$.

Proof. If $\rho_{ji} < 1$, $P_i(X_{\tau_j+m} \neq i\ \forall\ m \ge 0) > 0$. Together with the fact that $i \to j$, this implies that $\rho_{ii} < 1$, that is, $i \notin S_R$, a contradiction. This implies that $\rho_{ji} = 1$. Let n_1, n_2 be integers satisfying $p^{n_1}(j, i) > 0$, $p^{n_2}(i, j) > 0$. Then $\alpha = p^{n_1}(j, i)\, p^{n_2}(i,j) > 0$. Note that

$$p^{n_1+n+n_2}(j, j) \ge \alpha\, p^n(i, i).$$

Summing over n and using the fact that $G(i,i) = \infty$, we have $G(j, j) = \infty$ and thus $j \in S_R$. By the symmetry of the first step, $\rho_{ij} = 1$. QED.

It follows that on S_R, $i \to j$ is an equivalence relation. $E \subset S$ is said to be closed if $P_i(E) = 1$ for any $i \in E$. It is said to be irreducible if it is closed, and furthermore $i \to j$ for $i, j \in E$. The Markov chain itself is said to be irreducible if S is. By the above theorem, it is then either transient (i.e., $S = S_T$) or recurrent (i.e., $S = S_R$).

Define $N_n(j) = \sum_{m=1}^{n} I\{X_m = j\}$, $G_n(i, j) = E_i[N_n(j)]$, $T_j = E_j[\tau_j]$ (possibly $+\infty$) and $\tau_j^n = R_j^n - R_j^{n-1}$ for $n \ge 1$ where $\{R_j^n\}$ are defined recursively as

$$R_j^0 = \tau_j,\ R_j^n = \min\{n > R_j^{n-1} \mid X_n = j\},\ n \ge 1.$$

Theorem 2.3. (a) $j \in S_T$ implies: $N_n(j)/n \to 0$, P_i-a.s. and $G_n(i,j)/n \to 0$ for all $i \in S$.
(b) $j \in S_R$ implies: $N_n(j)/n \to T_j^{-1} I\{\tau_j < \infty\}$ P_i-a.s. and $G_n(i,j)/n \to \rho_{ij} T_j^{-1}$ where by convention, $a/\infty = 0$ for any $a \in R$.

Proof. (a) Since $N_n(j) \le N_j < \infty$ P_i-a.s. and $G_n(i,j) \le G(i,j) < \infty$, the claim follows immediately.
(b) Let $i = j$. By SMP, $\{\tau_j^n, n \ge 1\}$ are i.i.d. By the strong law of large numbers,

$$R_j^n/n = n^{-1} \sum_{m=1}^{n} \tau_j^m \to T_j \quad P_j\text{-a.s.}$$

Since $R_j^{N_n(j)} \le n < R_j^{N_n(j)+1}$, the first claim follows for $i = j$. For $i \ne j$, apply the same argument to the Markov chain $\{X_{\tau_j+m}, m \ge 0\}$ defined on the probability space (Ω', F', P') where $\Omega' = \{\tau_j < \infty\}$. F' = the corresponding trace σ-field and $P' = P$ restricted to $\{\tau_j < \infty\}$ and renormalized to make it a probability measure. (Of course, we assume that $P(\tau_j < \infty) > 0$, the other case being trivial.) The first half of (b) follows. The claim concerning $G_n(i,j)$ now follows by the dominated convergence theorem. QED.

Let $S_P = \{j \in S_R \mid T_j < \infty\}$ and $S_N = \{j \in S_R \mid T_j = \infty\}$. The elements of S_P and S_N are known respectively as the positive recurrent and the null recurrent states.

Theorem 2.4. If $i \in S_p$ and $i \to j$, then $j \in S_p$.

Proof. Since $i \to j$, $j \to i$. Let n_1, n_2 be integers such that $p^{n_1}(j,i) > 0$, $p^{n_2}(i,j) > 0$. Then for

$$\alpha = p^{n_1}(j,i)p^{n_2}(i,j) > 0.$$

we have

$$p^{n_1+n+n_2}(j,j) \ge \alpha\, p^n(i,i).$$

Hence

$$\frac{1}{n} \sum_{m=1}^{n} p^{n_1+n+n_2}(j,j) = \frac{G_{n_1+n+n_2}(j,j)}{n} - \frac{G_{n_1+n_2}(j,j)}{n}$$

$$\ge \alpha\, G_n(i,i)/n.$$

Letting $n \to \infty$, the left-hand side and the right-hand side converge respectively to T_j^{-1} and

$\alpha\ T_i^{-1}$, implying $T_j^{-1} > 0$, that is, $T_j < \infty$. QED.

It follows that $i \to j$ is an equivalence relation on S_p and S_N. Thus S can be written as the disjoint union $S = S_T \cup S_p \cup S_N$, where each $i \in S_p$ or S_N lies in an equivalence class with respect to the above equivalence relation. Each such class, called a communicating class, is obviously closed. Also, it is clear that $p^n(i,j) = 0$, $n \geq 0$, when i,j belong to two distinct equivalence classes of the above type and thus

$$n^{-1} \sum_j G_n(i, j) = 1, \quad n \geq 1, \tag{2.1}$$

where j is summed over such an equivalence class containing i. If this equivalence class is finite, Theorems 2.3(b) and (2.1) lead to

$$\sum_j T_j^{-1} = 1$$

where the summation is as above. Thus at least one of the T'_js is finite, implying that the corresponding j is in S_p. Thus the equivalence class itself must be in S_p if it is finite. If it is a singleton, say $\{i\}$, i is called an absorbing state and must satisfy

$$p(i, j) = \delta_{ij} \text{ (the Kronecker delta)}, \; j \in S.$$

Combining Theorems 2.1-2.4 above, we can now describe the qualitative behaviour of a general Markov chain. A Markov chain starting at $i \in S_R$ remains in the communicating class containing i, and visits each state in this class infinitely often with probability one. The mean return time to state i, $E_i[\tau_i]$, is finite if $i \in S_p$ and infinite if $i \in S_N$. If it starts at some $i \in S_T$, then there are two possibilities. One is that it can end up in a finite (random) time in one of the equivalence classes of S_R described above and remain in it thereafter. The second possibility is that it never hits S_R, but drifts to infinity. That is, for each finite subset of S, there is a finite random time after which the chain does not visit this set. If S_T is finite, only the first possibility exists.

The Markov chain can be represented by a weighted digraph (directed graph) as follows: Let each $i \in S$ correspond to a node of the graph, and draw an edge directed from i to j if $p(i, j) > 0$. Assign a weight $p(i, j)$ to this edge. Thus at each node, weights of the outgoing edges add up to one. A node corresponds to an absorbing state if and only if the only outgoing edge from the node returns to itself. The relation $i \sim j$ on the nodes defined by 'a directed path exists from i to j and from j to i' is an equivalence relation, and corresponds to $i \to j$, $j \to i$ above. A closed equivalence class under this definition will be called a communicating class. This will correspond to the concept of a communicating class defined earlier if it lies in S_R. Note however that with this new definition, a communicating class can also be transient. Graphically, a communicating class is characterized as follows. A is a communicating class if no edges leave A and for any $i, j \in A$, directed paths exist from i to j and from j to i. Note also that recurrence and positive recurrence of a communicating class of infinite cardinality do not depend only on its directed graph, but also depend on the weights we assign to its edges. Finally, the chain is irreducible if S is a single communicating class.

II.3 Stationary Distributions

A measure π on S is said to be stationary or invariant for the Markov chain $\{X_n\}$ if

$$\sum_{j \in S} p(j, k)\pi(j) = \pi(k), \quad k \in S,$$

where $\pi(i) = \pi(\{i\})$ for $i \in S$. Writing π as a row vector $[\pi(1), \pi(2),...]$, the above becomes $\pi P = \pi$ in matrix notation. Thus $\pi P^n = \pi$ for $n \geq 1$ and hence $\pi G_n/n = \pi$, $n \geq 1$, where $G_n = [\, [G_n(i, j)]\,]$. If π is a probability measure in addition, we call it a stationary distribution or an invariant probability measure. It is clear that if π is an invariant probability measure and the law of X_0 is π, then the law of X_n will be π for all n. In fact, the laws of $(X_0, X_1, X_2,...)$ and $(X_n, X_{n+1}, X_{n+2},...)$ agree for all n, making $\{X_n\}$ a stationary process.

Theorem 3.1. If π is an invariant probability measure for $\{X_n\}$, then $\pi(i) = 0$ for all $i \notin S_p$.

Proof. For $i \notin S_p$, $G_n(j, i)/n \to 0$. By the dominated convergence theorem, the i'th component of $\pi G_n/n$ tends to zero as $n \to \infty$. But $\pi G_n/n = \pi$. The claim follows. QED.

Theorem 3.2. Let $\{X_n\}$ be irreducible and positive recurrent (that is, $S = S_p$). Then there exists a unique stationary distribution $\pi \in \mathbb{P}(S)$ given by

$$\pi(i) = T_i^{-1}, \ i \in S.$$

Proof. If π is a stationary distribution, then

$$G_n(i,j)/n \to T_j^{-1}$$

and

$$\pi = \pi G_n/n$$

together imply (by the dominated convergence theorem) that $\pi(i) = T_i^{-1}$, $i \in S$. Thus it suffices to show that $\pi(i) = T_i^{-1}$, $i \in S$, defines a stationary distribution. Let S be finite. Since

$$\sum_{j \in S} G_n(i, j)/n = 1,$$

one has

$$1 = \lim_{n \to \infty} \sum_{j \in S} G_n(i, j)/n = \sum_{j \in S} \lim_{n \to \infty} G_n(i, j)/n = \sum_{j \in S} T_j^{-1}.$$

Since $G_n P = G_{n+1} - P$,

$$\begin{aligned} T_j^{-1} &= \lim_{n \to \infty} G_n(i, j)/n \\ &= \lim_{n \to \infty} \sum_k G_n(i, k)p(k, j)/n \\ &= \sum_{k \in S} \lim_{n \to \infty} G_n(i, k)p(k, j)/n \\ &= \sum_{k \in S} T_k^{-1} p(k, j), \end{aligned} \tag{3.1}$$

for $j \in S$, proving the desired result. If S is infinite, take $S_1 \subset S$ finite. Let $n \to \infty$ in the inequality

$$\sum_{j \in S_1} G_n(i, j)/n \leq 1$$

to conclude that

$$\sum_{j \in S_1} T_j^{-1} \leq 1.$$

Similarly, we argue as for (3.1) to conclude

$$\sum_{j \in S_1} T_j^{-1} p(j, k) \leq T_k^{-1}, \quad k \in S.$$

Thus

$$\sum_{j \in S} T_j^{-1} \overset{\Delta}{=} c^{-1} \leq 1 \tag{3.2}$$

$$\sum_{j \in S} T_j^{-1} p(j, k) \leq T_k^{-1}, \; k \in S. \tag{3.3}$$

Summing (3.3) over k, we get,

$$\sum_{j \in S} T_j^{-1} \leq \sum_{k \in S} T_k^{-1}.$$

Thus we must have equality in (3.3), implying that $\pi(i) = cT_i^{-1}$, $i \in S$, defines a stationary distribution. From the first part of the proof, it follows that $c = 1$. QED.

Let $\Gamma \subset \mathbb{P}(S)$ denote the set of stationary distributions for $\{X_n\}$. If $S_p = \emptyset$, $\Gamma = \emptyset$ by Theorem 3.1. If $S_p \neq \emptyset$, support $(\pi) \subset S_p$ for all $\pi \in \Gamma$. Let $C_1, C_2, \ldots$ be the (possibly infinitely many) communicating classes in S_p. They are mutually disjoint. Let $P^{(i)} = [[p(j, k)]]$ for $j, k \in C_i$. Then $P^{(i)}$ is a transition matrix on C_i. A Markov chain with state space C_i and transition matrix $P^{(i)}$ will be irreducible positive recurrent. By the above theorem, there is a unique $\pi_i \in \mathbb{P}(C_i)$ such that $\pi_i P^{(i)} = \pi_i$, $i = 1,2,\ldots$. We may consider π_i as an element of $\mathbb{P}(S)$ by setting $\pi_i(j) = 0$ for $j \notin C_i$. On the other hand for any $\pi \in \Gamma$ such that

$$a_i \overset{\Delta}{=} \sum_{j \in C_i} \pi(j) > 0$$

the measure $\tilde{\pi}_i \in \mathbb{P}(C_i)$ defined by

$$\tilde{\pi}_i(j) = \pi(j)/a_i, \; j \in C_i$$

satisfies $\tilde{\pi}_i P^{(i)} = \tilde{\pi}_i$ and hence coincides with π_i. Thus any $\pi \in \Gamma$ may be written as $\pi = \Sigma a_i \pi_i$ where

$$a_i = \sum_{j \in C_i} \pi(j), \; i = 1,2,\ldots ,$$

satisfy

$$a_i \geq 0, \sum_i a_i = 1.$$

Conversely, any convex combination of $\{\pi_i\}$ is in Γ, as is easily verified. Thus we have:

Corollary 3.1. $\Gamma = \{\pi \in \mathbb{P}(S) \mid \pi = \sum_i a_i\pi_i, a_i \geq 0, \sum_i a_i = 1\}$, where $\{\pi_i\}$ are as above, viewed as elements of $\mathbb{P}(S)$ by assigning zero mass to the complement of the corresponding C_i.

For $i \in S$ satisfying $\rho_{ii} > 0$ (i.e., $p^n(i, i) > 0$ for some $n \geq 1$), define the period of i, denoted d_i, to be the greatest common divisor of the set $\{n > 1 \mid p^n(i, i) > 0\}$.

Lemma 3.1. If $i \to j$ and $j \to i$, $d_i = d_j$.

Proof. Take $n_1 \geq 1$, $n_2 \geq 1$ such that

$$p^{n_1}(i, j) > 0, \quad p^{n_2}(j, i) > 0.$$

Thus

$$p^{n_1+n_2}(i, i) > 0.$$

Therefore d_i divides $n_1 + n_2$. For any n satisfying

$$p^n(j, j) > 0,$$

we have

$$p^{n_1+n+n_2}(i, i) > 0,$$

implying that d_i divides $n_1 + n + n_2$. Thus d_i must divide n, implying $d_i \leq d_j$. A symmetric argument gives $d_j \leq d_i$. QED.

An irreducible Markov chain is said to be periodic with period d if $d_i = d > 1$ for all i, and aperiodic if $d_i = 1$ for all i. (Recall from the above lemma that d_i will be independent of i.)

Theorem 3.3. Let $\{X_n\}$ be irreducible and positive recurrent with stationary distribution π.

(a) If $\{X_n\}$ is aperiodic, $\lim_{n\to\infty} p^n(i, j) = \pi(j)\ \forall\, i, j \in S$.

(b) If $\{X_n\}$ is periodic with period d, then for each pair (i, j) in $S \times S$, there exists an interger r with $0 \le r < d$ such that $p^n(i, j) = 0$, unless $n = md + r$ for some $m \ge 0$ and

$$\lim_{m\to\infty} p^{md+r}(i, j) = d\pi(j).$$

The proof will be based on the following lemma.

Lemma 3.2. Let $I \subset \mathbb{N}_0 = \{0,1,2,...\}$ satisfy: (i) $I + I \subset I$, (ii) the greatest common divisor of I is 1. Then there exists an $n_0 \in \mathbb{N}$ such that $n \in I$ for $n \ge n_0$.

Proof. We claim that there exists an $n_1 \in \mathbb{N}_0$ such that both n_1 and $n_1 + 1$, are in I. If not, there exists $k \ge 2$ in $\mathbb{N}_0$ and $n_1 \in I$ such that $n_1 + k \in I$ and for all a, b in I with $a \ne b$, $|a - b| \ge k$. Since the greatest common divisor of I is 1, there exists $n \in I$ such that k does not divide n. Thus $n = mk + r$ for some $m \in \mathbb{N}_0$, $0 < r < k$. Since $I + I \subset I$, $(m+1)(n_1+k)$ and $n+(m+1)n_1$ are in I. Their difference is $k - r < k$, a contradiction. Thus the claim holds. Now take $n_0 = n_1^2$. For $n \ge n_1^2$, $n - n_1^2 = mn_1 + r$ for some $m \in \mathbb{N}_0$, $\theta \le r < n_1$. Thus $n = n_1^2 + mn_1 + r = (n_1 + 1)r + (n_1 - r + m)n_1 \in I$. QED

Proof of the Theorem: (a) Let $s \in S$ and $I = \{n \ge 1 \mid p^n(s, s) \ge 0\}$. Then the greatest common divisor of I is 1 and $I + I \subset I$. So there exists an $n_1 \in \mathbb{N}_0$ such that $n \in I$ for $n \ge n_1$, i.e., $p^n(s,s) > 0$ for all $n \ge n_1$. Take n_2, n_3 in $\mathbb{N}$ such that $p^{n_2}(i, s) > 0$ and $p^{n_3}(s, j) > 0$. Then for $n_0 = n_1 + n_2 + n_3$ (which depends on i, j), we have $p^n(i, j) > 0$ for all $n \ge n_0$. Consider an $S \times S$-valued Markov chain $\{(X_n, Y_n)\}$ with the transition matrix $\bar{P}$ given $[[\bar{p}((i, j), (k, \ell))]]$, $(i, j), (k, \ell) \in S \times S$, where,

$$\bar{p}((i, j),\ (k, \ell)) = p(i, k)p(j, \ell).$$

From the foregoing, it follows that this chain is irreducible and aperiodic. It is trivial to verify that $\bar{\pi} \in \mathbb{P}(S \times S)$, defined by $\bar{\pi}((i, j)) = \pi(i)\pi(j)$ for $i, j \in S$, is a stationary distribution for $\{(X_n, Y_n)\}$. Thus $\{(X_n, Y_n)\}$ is positive recurrent. Let $\tau = \min\{n \ge 1 | X_n = Y_n\}$. Since $\tau \le \tau_{(i,i)}$ for any $i \in S$ and $(i,i) \in (S \times S)_p$, we have $\tau < \infty$ a.s. By the SMP,

$$P(X_n = j,\ \tau \le n) = P(Y_n = j,\ \tau \le n)$$

for all $j \in S$ and any law of (X_0, Y_0). Hence

$$P(X_n = j) = P(X_n = j,\ \tau \le n) + P(X_n = j,\ \tau > n)$$

$$= P(Y_n = j,\ \tau \le n) + P(X_n = j,\ \tau > n)$$

$$\le P(Y_n = j) + P(\tau > n).$$

By a symmetrical argument,

$$P(Y_n = j) \le P(X_n = j) + P(\tau > n).$$

Thus

$$| P(X_n = j) - P(Y_n = j) | \le P(\tau > n) \to 0$$

as $n \to \infty$. Take $P(X_0 = i) = 1$, $P(Y_0 = k) = \pi(k)$ for $k \in S$ to conclude the proof.

(b) First we extend (a) slightly. Suppose we drop the assumption of irreducibility and assume that $S_p \neq \emptyset$. If $C \subset S_p$ is closed and irreducible, then there is a unique stationary distribution π supported on C. By considering the Markov chain restricted to C, we find that $p^n(i, j) \to \pi(j) = T_j^{-1}$ for all $(i, j) \in C \times C$. Coming back to (b), let $Y_n = X_{nd}$, $n \ge 1$. It is easily verified that $\{Y_n\}$ is an aperiodic Markov chain with transition matrix $Q = P^d$, written as $[\,[q(i,j)]\,]$. Write $Q^n = [\,[q^n(i, j)]\,]$, $n \ge 1$. Let $\bar{\tau}_j = \min\{n \ge 1 \mid Y_n = j\}$. Then $E_j[\bar{\tau}_j] = d^{-1} E_j[\tau_j] = d^{-1} T_j < \infty$. Hence j is positive recurrent for $\{Y_n\}$. Letting $C = \{i \in S \mid i \to j \text{ for } \{Y_n\}\}$, the foregoing discussion implies that $q^m(i,j) = p^{md}(i, j) \to dT_j^{-1}$ $= d\pi(j)$. Fix (i, j) and let $r_1 = \min\{n \ge 1 \mid p^n(i, j) > 0\}$. Then $p^{r_1}(i, j) > 0$. If $p^m(j, i) > 0$ for some m, then $p^{r_1+m}(j, j) > 0$. More generally, for any n satisfying $p^n(i, j) > 0$, $p^{n+m}(j, j) > 0$. Hence d divides $m + r_1$ and $m + n$ and therefore, $n - r_1$. Write $r_1 = m_1 d + r$ with $0 \le r < d$, $m_1 \in \mathbb{N}$. Then $n = md + r$ for some $m \in \mathbb{N}_0$. Since by the SMP,

$$p^n(i, j) = \sum_{k=1}^{n} P_i(\tau_j = k)\, p^{n-k}(j, j),\ \ n \ge 1,$$

we have

$$p^{md+r}(i, j) = \sum_{k=0}^{m} P_i(\tau_j = kd + r) p^{(m-k)d}(j, j)$$

$$= \sum_{k=0}^{\infty} P_i(\tau_j = kd + r) p^{(m-k)d}(j, j) I\{k \le m\}.$$

As $m \to \infty$,

$$p^{(m-k)d}(j, j) \to d\pi(j).$$

The dominated convergence theorem now leads to

$$p^{md+r}(i, j) \to d\pi(j)P_i(\tau_j < \infty) = d\pi(j).$$

This completes the proof. QED

Corollary 3.2 Suppose $\{X_n\}$ is irreducible and either transient or null recurrent. Then $p^n(i, j) \to 0$ as $n \to \infty$ for all $i, j \in S$.

Proof. The transient case is immediate from the fact $G(i,j) < \infty$ for all i,j. For the null recurrent case, assume without any loss of generality that the chain is aperiodic. Let P^* be a limit point of $\{P^n\}$ in $[0, 1]^{\infty \times \infty}$, say

$$P^{n(k)} \to P^*$$

termwise. (This is possible by the compactness of $[0, 1]^{\infty \times \infty}$.) Argue as in part (a) of the above theorem to conclude that

$$|p^n(i, j) - p^n(k, j)| \to 0 \text{ as } n \to \infty, \ i,j, k \in S.$$

Hence all rows of P^* are identical. Thus $PP^* = P^*$. By the dominated convergence theorem,

$$P^{n(k)+1} = PP^{n(k)} \to PP^* = P^*.$$

By Fatou's lemma,

$$P^*P = (\lim_{k \to \infty} P^{n(k)})P \le \lim(P^{n(k)}P) = P^*$$

where the inequality is termwise. Letting μ denote any row of P^*, $\mu(i) \in [0, 1]$ for all i and

$$\mu_0 \overset{\Delta}{=} \sum_{i \in S} \mu(i) \le 1$$

by Fatou's lemma. If $\mu_0 > 0$, let $\tilde{\mu}(i) = \mu_0^{-1}\mu(i)$ for $i \in S$. Then $\tilde{\mu}(i) \in [0, 1]$ for all i,

$\sum_i \tilde{\mu}(i) = 1$ and

$$\tilde{\mu}P \leq \tilde{\mu}$$

in view of the foregoing. But

$$1 = \sum_i (\tilde{\mu}P)(i) \leq \sum_i \tilde{\mu}(i) = 1.$$

Hence $\tilde{\mu}P = \tilde{\mu}$, i.e., $\tilde{\mu}$ is a stationary distribution for $\{X_n\}$. This contradicts the fact that $\{X_n\}$ is not positive recurrent. Thus $\mu_0 = 0$ and the claim follows. QED

Theorem 3.4. Let $\{X_n\}$ be positive recurrent and irreducible with π its invariant probability measure. Then π is given by

$$\int f d\pi = E\Big[\sum_{m=0}^{\tau_1 - 1} f(X_m)/X_0 = 1\Big]/E\,[\tau_1/X_0 = 1] \qquad (3.4)$$

for $f \in C_b(S)$.

Proof. Let $X_0 = 1$ and $\{\sigma_i\}$ the successive return times to state 1 with $\sigma_0 = 0$. By SMP, the law of $(X_{\sigma_n}, X_{\sigma_n+1},\ldots)$ coincides with its conditional law given F_{σ_n}, and in turn with the law of $(X_0, X_1,\ldots)$ for all $n \geq 0$. It follows that for any $f \in C_b(S)$,

$$\sum_{m=\sigma_n}^{\sigma_{n+1}-1} f(X_m), \quad n \geq 0,$$

are i.i.d. Now

$$\Big(\sum_{m=0}^{\sigma_n - 1} f(X_n)\Big)/\sigma_n = \frac{\Big(\sum_{i=0}^{n-1}\Big(\sum_{m=\sigma_i}^{\sigma_{i+1}-1} f(X_m)\Big)\Big)/n}{\Big(\sum_{i=0}^{n-1}(\sigma_{i+1} - \sigma_i)\Big)/n}$$

The left-hand side a.s. tends to the left hand side of (3.4) by Theorem 2.3(b) and Theorem 3.2. The right-hand side converges a.s. to the right-hand side of (3.4) by the strong law of large numbers. QED

In conclusion, we remark that the theory developed so far deals with Markov chains with stationary transitions, that is, Markov chains whose transitions between any two prescribed elements of S are governed by the corresponding element of a fixed transition matrix P, regardless of when the transition is to take place. They are also called time-homogeneous Markov chains. More generally, one can consider 'time-inhomogeneous Markov chains' whose transitions are governed by a sequence of transition matrices $P_n = [[p_n(i, j)]]$, $i, j \in S$, $n \geq 0$, rather than by a single transition matrix P. Thus $\{P_n\}$ satisfy:

$$p_n(i, j) \in [0, 1], \ \sum_{j \in S} p_n(i,j) = 1, \ n \geq 0, \ i, j \in S$$

and (c) of the first section is replaced by

$$P(X_{n+1} = j / X_n, X_{n-1}, \ldots, X_0) = p_n(X_n, j), \ n \geq 0.$$

Then $\{X_n\}$ continues to be a Markov process. The SMP also holds, as arguments analogous to those for the time-homogeneous case show. However, a lot of the foregoing theory does not go over for obvious reasons. See [IsMa] for some results on time-inhomogeneous Markov chains.

Chapter II
Controlled Markov Chains

This chapter recalls the notation of [Bor 1]-[BoGh] in the first section, and introduces typical control problems in the second section along the lines of [Bor 7]. The last section derives some general results about controlled Markov chains. These are from [Bor 4] and [Bor 5].

II.1 Notation and Terminology

Let $S = \{1,2,...\}$ as before. From now on, X_n, $n \geq 0$, will denote a controlled Markov chain on state space S. That is, X_n, $n \geq 0$, is an S-valued process whose transitions follow the mechanism described in the last chapter except for an additional 'control' parameter that enters the transition probabilities. This is to be chosen by a decision maker (or 'controller') in the background, based on the information available to him. This section describes the precise formulation of this mechanism.

Instead of a single transition matrix, we now consider a family of transition matrices $P_u = [\,[p(i,j,u_i)]\,]$, $i, j \in S$, indexed by the 'control vector' $u = [u_1 u_2,...]$. Here, $u_i \in D(i)$, $i \in S$, for some prescribed compact metric spaces $D(i)$, $i \in S$, called control spaces or action spaces. The functions $p(i,j,.)$, $i,j, \in S$, are assumed to be continuous. By replacing each $D(i)$ by $\prod_k D(k)$ and $p(i,j,.)$ by its composition with the projection $\prod_k D(k) \to D(i)$ for each i, j, one may assume that all $D(i)$'s are in fact replicas of a fixed compact metric space D. (This statement will be qualified in our study of control under partial observations.)

Next we introduce some general notation, some of which we have already been using. For any Polish space Y, Y^n, $n = 1,2,...,\infty$, will denote the n-times product of copies of Y with the product topology. We have already introduced the space $\mathbb{P}(Y)$ of probability measures on Y wth the Prohorov topology. Similarly, let $M(Y)$ denote the space of finite non-negative measures on Y with topology of weak convergence, i.e., the coarsest topology that makes the maps $\mu \in M(Y) \to \int f d\mu \in R$ continuous for all $f \in C_b(Y)$ (= the space of bounded continuous functions $Y \to R$ with the supremum norm). Then $\mathbb{P}(Y) \subset M(Y)$ with the relative topology. When Y is a finite or a countable set, say $\{1,2,...\}$, we can write $\nu \in M(Y)$ as a row vector $[\nu(\{1\}), \nu(\{2\}),...,]$, also written as $[\nu(1), \nu(2),...]$, as was done with $\pi \in \mathbb{P}(S)$ in section I.3. We denote this row vector by ν again by abuse of notation. ν will refer to the measure or the vector according to context.

Let $L = D^\infty$. A control strategy (CS) is a sequence $\{\xi_n\}$, $\xi_n = [\xi_n(1), \xi_n(2),...]$, $n \geq 0$, of L-valued random variables such that for $i \in S$, $n \geq 0$ and $\mathcal{F}_n$ = the σ-field generated by X_m, ξ_m, $m \leq n$,

$$P(X_{n+1} = i/\mathcal{F}_n) = p(X_n, i, \xi_n(X_n)). \tag{1.1}$$

$\{X_n\}$ is said to be governed by the CS $\{\xi_n\}$ whenever (1.1) holds. If ξ_n is independent of X_m, $m \le n$, ξ_m, $m < n$, for each n, we call the control strategy $\{\xi_n\}$ a Markov randomized strategy (MRS). This is because under such a CS, $\{X_n\}$ is a Markov chain, though not necessarily one with stationary transitions. In fact, the transition probabilities are explicitly given as follows.

Define

$$p_\Phi(i, j) = \int p(i,j,u)\ \hat{\Phi}_i(du)$$

for $i,j \in S$, $\Phi \in \mathbb{P}(L)$ with $\hat{\Phi}_i \in P(D)$ being the image of Φ under the projection map from L to its i-th factor space. Let $P[\Phi]$ denote the stochastic matrix $[[p_\Phi(i, j)]]$. If $\{\xi_n\}$ is an MRS and the law of ξ_n is Φ_n for each n, one has

$$\begin{aligned} P(X_{n+1} = i/X_m,\ m \le n) \\ &= E[p(X_n, i, \xi_n(X_n))/X_m,\ m \le n] \\ &= p_{\Phi_n}(X_n, i),\ n \ge 0. \end{aligned}$$

Thus under the CS $\{\xi_n\}$, $\{X_n\}$ is a Markov chain governed by the transition matrices $\{P[\Phi_n]\}$.

Note that the transition matrices depend on $\{\Phi_n\}$ only through the marginal distributions thereof on the factor spaces of L. In the control problems we are going to consider, we shall be interested only in the laws of the sequence $\{(X_n, \xi_n(X_n)),\ n \ge 0\}$ and not of $\{(X_n, \xi_n),\ n \ge 0\}$. Thus for an MRS as above, we may assume that Φ_n, $n \ge 0$, lie in $\mathbb{P}_0(L) \overset{\Delta}{=}$ the space of product probability measures on L, viewed as a subset of $\mathbb{P}(L)$ with the relative topology. Note that $\mathbb{P}_0(L)$ is compact in $\mathbb{P}(L)$. We shall write a typical $\Phi \in \mathbb{P}_0(L)$ as $\Phi = \prod_i \hat{\Phi}_i$ with $\hat{\Phi}_i \in \mathbb{P}(D)$.

Suppose an MRS $\{\xi_n\}$ satisfies in addition the condition that $\{\xi_n\}$ are identically distributed with a common law, say, $\Phi \in \mathbb{P}_0(L)$. The MRS is then said to be a stationary randomized strategy (SRS), denoted by $\gamma[\Phi]$. The corresponding process $\{X_n\}$ will be a Markov chain with stationary transitions governed by the transition matrix $P[\Phi]$.

One can specialize even further. Suppose the law of each ξ_n in an MRS $\{\xi_n\}$ is a Dirac measure. We call the MRS simply a Markov strategy (MS). Similarly, if Φ is a Dirac measure at, say, $\xi \in L$, call the SRS $\gamma[\Phi]$ a stationary strategy (SS), to be denoted by $\gamma\{\xi\}$. The corresponding transition matrix will now be denoted by $P\{\xi\} = P_\xi$. Observe that the above nomenclature corresponds, as it should, to one's intuitive concepts of randomized or non-randomized controls, as applicable.

Say that an SRS $\gamma[\Phi]$ or an SS $\gamma\{\xi\}$ is a stable SRS (SSRS) or a stable SS (SSS) respectively if the corresponding Markov chain is positive recurrent. If in addition it has a single communicating class, it will also have a unique invariant probability measure, as described in section I.3. We denote this measure by $\pi[\Phi]$, $\pi\{\xi\}$ respectively.

The stage is now set for formulating the control problems, which we consider in the next section.

II.2 The Control Problems

Let $h : S \to R^+$, $k : S \times D \to R^+$, $\iota : \{0, 1, ..., N\} \times S \times D \to R^+$ be continuous functions. The classical control problems that we shall be studying are the following:

(C1) Minimize over all CS

$$E[\sum_{n=0}^{\infty} \beta^n k(X_n, \xi_n(X_n))], \ 0 < \beta < 1. \tag{2.1}$$

This is the infinite horizon discounted cost control problem. The number β is called the discount factor and the function k the running cost.

(C2) Minimize over all CS

$$E[\sum_{n=0}^{T-1} k(X_n, \xi_n(X_n)) + h(X_T)] \tag{2.2}$$

where for a prescribed finite set A in S,

$$T = \min \{n \geq 0 \mid X_n \notin A\}. \tag{2.3}$$

This is called 'control up to the first exit time T' and k, h respectively the running cost and the terminal cost (or exit cost).

(C3) Minimize over all CS

$$E[\sum_{n=0}^{N-1} \iota(n, X_n, \xi_n(X_n)) + h(X_N)], \ 0 < N < \infty. \tag{2.4}$$

This is the 'finite horizon control problem' with ι, h its running cost and terminal cost respectively.

(C4) Minimize a.s. over all CS

$$\limsup_{n \to \infty} \frac{1}{n} \sum_{m=0}^{n-1} k(X_m, \xi_m(X_m)), \tag{2.5}$$

This is the 'ergodic' or 'long run average cost' control problem with 'running cost' k. Note that under an SSRS $\gamma[\Phi]$ with $\{X_n\}$ having a single communicating class, (2.5) a.s. equals

$$\sum_{i \in S} \pi[\Phi]\,(i) \int_S k(i, u)\, \hat{\Phi}_i(du) \tag{2.6}$$

To each of the above problems, one can naturally associate a suitably defined 'occupation measure' such that the cost functional in question can be described as the integral of some function with respect to this measure. We consider these one by one.

Consider the infinite horizon discounted cost control problem (C1). The discounted occupation measure for initial condition $X_0 = i$ is $\nu_{Di} \in M(S \times D)$ defined by

$$\int f d\nu_{Di} = E\Big[\sum_{n=0}^{\infty} \beta^n f(X_n, \xi_n(X_n))/X_0 = i\Big] \tag{2.7}$$

for $f \in C_b(S \times D)$, $i \in S$. For $\pi \in \mathbb{P}(S)$, the discounted occupation measure $\nu_{D\pi} \in M(S \times D)$ for initial law π will be defined by

$$\int f d\nu_{D\pi} = \sum_{i \in S} \pi(\{i\}) \int f d\nu_{Di} \tag{2.8}$$

for $f \in C_b(S \times D)$. The cost for (C1) can now be expressed as

$$\int k\, d\nu_{D\pi} \tag{2.9}$$

where π is the law of X_0.

For the problem of control up to the first exit time T, we anticipate a result proved in Chapter IV and assert that under suitable hypotheses, $E[T]$ is finite under any CS and any initial law. This allows us to define the 'occupation measure up to T' for the initial condition $X_0 = i$ as $\nu_{Ti} \in M(A \times D)$ defined by

$$\int f d\nu_{Ti} = E\Big[\sum_{n=0}^{T-1} f(X_n, \xi_n(X_n))/X_0 = i\Big] \tag{2.10}$$

for $f \in C(A \times D)$. We consider this problem only for initial laws supported in A. For $\pi \in \mathbb{P}(A)$, define $\nu_{T\pi} \in M(A \times D)$ by

$$\int f d\nu_{T\pi} = \sum_{i \in A} \pi(\{i\}) \int f d\nu_{Ti}. \tag{2.11}$$

Now define $\bar{k} \in C(A \times D)$ by

$$\bar{k}(i, u) = k(i, u) + \sum_{j \in S} p(i,j,u)\, h(j) - h(i) \tag{2.12}$$

Under any CS $\{\xi_n\}$, the sequence $(M_n, \mathcal{F}_n)$ with

$$M_n = \sum_{m=1}^{n-1} [\, k(X_m, \xi_m(X_m)) + h(X_{m+1}) - h(X_m) - \bar{k}\,(X_m, \xi_m(X_m))] \tag{2.13}$$

is a zero mean martingale. An application of the optional sampling theorem to $(M_n, \mathcal{F}_n)$ and the definition of $\bar{k}$ lead to the conclusion that (2.2) equals

$$\int \bar{k}\, d\nu_{T\pi} - \int h d\pi, \tag{2.14}$$

π being the law of X_0. We may set $h \equiv 0$ on A and drop the second term.

For the finite horizon control problem (C3), the finite horizon occupation measure for initial condition $X_0 = i$ is given by $\nu_{Fi} \in M(\{0, 1,2,...,N\} \times S \times D)$, defined by

$$\int f d\nu_{Fi} = E[\, \sum_{m=0}^{N} f(m, X_m, \xi_m(X_m))/X_0 = i\,] \tag{2.15}$$

for $f \in C_b(\{0, 1,...,N\} \times S \times D)$. For a general initial law $\pi \in \mathbb{P}(S)$, we correspondingly define the occupation measure $\nu_{F\pi}$ by

$$\int f d\nu_{F\pi} = \sum_{i \in S} \pi(\{i\}) \int f d\nu_{Fi} \tag{2.16}$$

for f as above. Define $\bar{\iota} \in C_b(\{0, 1,...,N\} \times S \times D)$ by

$$\bar{\iota}(m, i, u) = \begin{cases} \iota\,(m, i, u) & 0 \le m < n \\ h(i) & m = N \end{cases} \tag{2.17}$$

The cost (2.4) of (C3) now becomes

$$\int \bar{\iota}\, d\nu_{F\pi}, \tag{2.18}$$

π being the law of X_0.

For the ergodic control problem, the ergodic occupation measure $\nu_E[\Phi] \in \mathbb{P}(S \times D)$ under an SSRS $\gamma[\Phi]$, $\Phi = \prod_i \hat{\Phi}_i$, is given by

$$\int f d\nu_E[\Phi] = \sum_{i \in S} \pi[\Phi](i) \int f(i, u)\, \hat{\Phi}_i(du) \tag{2.19}$$

for $f \in C_b(S \times D)$. Write $\nu_E\{\xi\}$ for $\nu_E[\Phi]$ if $\gamma[\Phi] = \gamma\{\xi\}$. Note that (2.6) can be expressed as

$$\int k d\nu_E[\Phi]. \tag{2.20}$$

We do not define an ergodic occupation measure under an arbitrary CS. There are obvious difficulties in doing so. However, (2.19) will suffice for our purposes, as we observe later in Chapter V.

In view of the foregoing, each of these control problems can be viewed as the problem of minimizing a linear functional on an appropriate set of measures. Our approach to these problems will be based on establishing the convexity of these sets of measures, thereby reducing the control problems to convex optimization problems. The next section paves the way for this programme by establishing some basic results concerning the laws of controlled Markov chains.

II.3 Laws of Controlled Markov Chains

This section establishes some broad facts regarding controlled Markov chains without reference to any specific cost criterion. The first of these concerns the conditional behaviour of controlled Markov chains conditioned on a stopped σ-field. Recall the family of σ-fields $\{\mathcal{F}_n\}$ defined above. Let τ be an $\{\mathcal{F}_n\}$-stopping time. For $n \geq 1$, let Z_n denote the Banach space obtained by completing with respect to the supremum norm the space of compactly supported continuous functions from $(S \times L)^n$ to R.

Theorem 3.1. On the set $\{\tau < \infty\}$, the regular conditonal law of the $(S \times L)$-valued process $(X_{\tau+n}, \xi_{\tau+n})$, $n \geq 0$, given $\mathcal{F}_\tau$ a.s. coincides with the law of some controlled Markov chain with transition matrices $\{P_u\}$, and its control process.

Proof. Let $(X, \xi)_\infty \overset{\Delta}{=} [(X_0, \xi_0), (X_1, \xi_1), \ldots]$ and let $F \in L_\infty(S \times L)^\infty)$, $B \in \mathcal{F}_\tau$. Then

$$E[E[F((X, \xi)_\infty)/\mathcal{F}_\tau]\, I_B\, I\{\tau < \infty\}]$$

$$= E[F((X, \xi)_\infty)\, I_B\, I\{\tau < \infty\}]$$

$$= \sum_{n=0}^{\infty} E[E[F((X, \xi)_\infty)/\mathcal{F}_n]\, I_B\, I\{\tau = n\}].$$

Since $B \in F_\tau$ was arbitrary, we have

$$E[F((X,\xi)_\infty)/\mathcal{F}_\tau] I\{\tau < \infty\} = \sum_{n=0}^{\infty} E[F((x,\xi)_\infty)/\mathcal{F}_n] I\{\tau = n\} \text{ a.s.} \tag{3.1}$$

For $m \geq 0$, $j \in S$ and $g \in Z_{m+1}$,

$$E[(I\{X_{\tau+m+1} = j\} - p(X_{\tau+m}, j, \xi_{\tau+m}(X_{\tau+m})))\, g((X_\tau, \xi_\tau), (X_{\tau+1}, \xi_{\tau+1}), \ldots, (X_{\tau+m}, \xi_{\tau+m}))/\mathcal{F}_\tau] = 0 \text{ a.s. on } \{\tau < \infty\} \tag{3.2}$$

by virtue of (3.1) and SMP, the former being applied to $F((X,\xi)_\infty) = I\{X_{\tau+m+1} = j\}$ with $\mathcal{F}_\tau$ replaced by $\mathcal{F}_{\tau+m}$. Let $N \in \mathcal{F}_\tau$ be the set of zero probability outside which (3.2) holds for all $m \geq 0$, $j \in S$ and g in a countable dense set of Z_{m+1}. For $w \in \Omega$, let $\mu(w)$ denote a version of the regular conditional law of $(X_{\tau+n}, \xi_{\tau+n})$, $n \geq 0$, given $\mathcal{F}_\tau$. Then the foregoing implies that for all w outside $N \cup \{\tau = \infty\}$, the following holds. Let (Y_n, ψ_n), $n \geq 0$, be an $(S \times L)$-valued process with law $\mu(w)$. Then for all $m \geq 0$,

$$E[(I\{Y_{m+1} = j\} - p(Y_m, j, \psi_m(Y_m)))\, g((Y_0, \psi_0), \ldots, (Y_m, \psi_m)))] = 0, \quad \mu(w)\text{ - a.s.}$$

for all g as above (and therefore for all $g \in Z_{m+1}$) and a suitable choice of $\mu(\cdot)$. A standard monotone class argument now leads to

$$E[I\{Y_{m+1} = j\}/Y_i, \psi_i, i \leq m] = p(Y_m, j, \psi_m(Y_m)) \quad \mu(w)\text{ - a.s.}$$

for all $m \geq 0$, $j \in S$. The claim follows. QED.

Let $\varphi_C, \varphi_{MR}, \varphi_{SR}, \varphi_M, \varphi_S \in \mathbb{P}((S \times D)^\infty)$ denote the sets of attainable laws of $[(X_1, \xi_1(x_1)), (X_2, \xi_2(X_2)), \ldots]$ as the control strategy varies over all CS, all MRS, all SRS, all MS and all SS respectively, the initial law being held fixed at some $\pi \in \mathbb{P}(S)$. For the sake of simplicity, we take π to be the Dirac measure at 1. The proofs of the following can be easily adapted for more general π. We now show that the above sets are compact and φ_c convex.

For $n \geq 0$ and any CS $\{\xi_m\}$, denote by $P^n(\{\xi_m\}, \cdot) \in \mathbb{P}(S)$ the law of X_n under $\{\xi_m\}$ with $X_0 = 1$. That is, $P^n(\{\xi_m\}, j) = P(X_n = j)$, $j \in S$, for $\{X_m\}$ governed by $\{\xi_m\}$ and starting at $X_0 = 1$.

Lemma 3.1. For each $n > 0$, the set of $P^n(\{\xi_m\}, \cdot)$ is tight in $\mathbb{P}(S)$ as $\{\xi_m\}$ varies over all CS.

Proof. We proceed by induction. The claim is trivial for $n = 0$. Suppose it holds for some $n \geq 0$. Let $\varepsilon > 0$. Pick $N \geq 1$ such that under all CS,

$$P(1 \leq X_i \leq N,\ 1 \leq i \leq n) > 1 - \varepsilon/2.$$

This is possible by the induction hypothesis. For each $i \in S$ and $u_n \to u$ in D, $p(i, j, u_n) \to p(i, j, u)$ for all $j \in S$. By Scheffe's theorem,

$$p(i,.,u_n) \to p(i,.,u)$$

in total variation and hence in $\mathbb{P}(S)$. Thus $p(i,.,u)$, $u \in D$, is tight in $\mathbb{P}(S)$. Pick $M_i \geq 1$ such that

$$\inf_u \sum_{j=1}^{M_i} p(i, j, u) \geq 1 - \varepsilon/2^{i+1},\ i = 1,2,\ldots,N.$$

Let $\overline{N} = \max\{M_1,\ldots,M_N, N\}$. A straightforward computation shows that under any CS,

$$\begin{aligned} &P(1 \leq X_i \leq \overline{N},\ 1 \leq i \leq n+1) \\ &= \sum_{j=1}^{\overline{N}} E[I\{1 \leq X_i \leq \overline{N},\ 1 \leq i \leq n\}\, E[X_{n+1} = j/\mathcal{F}_n]] \\ &\geq \sum_{j=1}^{\overline{N}} E[I\{1 \leq X_i \leq N,\ 1 \leq i \leq n\}\, p(X_n, j, \xi_n(X_n))] \\ &\geq (1 - \varepsilon/2)\, P(1 \leq X_i \leq N,\ 1 \leq i \leq n) > 1 - \varepsilon. \end{aligned}$$

The claim now follows by induction. QED

Theorem 3.2. φ_C is compact.

Proof. Let $\{X_n^m\}$, $m \geq 1$, be a sequence of controlled Markov chains governed by CS $\{\xi_n^m\}$, $m \geq 1$, respectively with $X_0^m = 1$ for all m. By the above lemma and compactness of L, it follows that the laws of (X_n^m, ξ_n^m), $m \geq 1$, are tight in $\mathbb{P}(S \times L)$ for each fixed n. Hence for $\xi^m = [\xi_0^m, \xi_1^m, \ldots]$ and $X^m = [X_0^m, X_1^m, \ldots]$, the laws of (ξ^m, X^m), $m \geq 1$, are tight in $\mathbb{P}(L^\infty \times S^\infty)$. Therefore they converge along a subsequence to the law of some $L^\infty \times S^\infty$-valued random variable (ξ^∞, X^∞). Dropping to this subsequence, denote it by $\{m\}$ again by abuse of notation. By Skorohod's theorem (see the appendix), we may assume that (ξ^m, X^m), m = 1,2,...∞, are defined on a common probability space and $(\xi^m, X^m) \to (\xi^\infty, X^\infty)$ a.s. in $L^\infty \times S^\infty$. Write $\xi^\infty = [\xi_0^\infty, \xi_1^\infty, ...]$ and $X^\infty = [X_0^\infty, X_1^\infty, ...]$. Let $n \geq 1, j \in S$ and $g \in Z_{n+1}$. Then the function $F : (S \times L)^{n+2} \to R$ defined by

$$F((x_0, u_0),...,(x_{n+1}, u_{n+1})) = (I\{x_{n+1} = j\} - p(x_n, j, u_n(x_n)))$$

$$g((x_0, u_0),...,(x_n, u_n))$$

is seen to be bounded and continuous. For m = 1,2,...,

$$E[(I\{X_{n+1}^m = j\} - p(X_n^m, j, \xi^m(X_n^m)))\, g((X_0^m, \xi_0^m),...,$$

$$(X_n^m, \xi_n^m))] = 0. \tag{3.3}$$

Letting $m \to \infty$, the continuity of F above implies that (3.3) holds for $m = \infty$ as well. Argue as in the proof of Theorem 3.1 to conclude that $\{X_n^\infty\}$ is a controlled Markov chain with transition matrices $\{P_u\}$, governed by $\{\xi_n^\infty\}$, with $X_0^\infty = 1$. The claim follows. QED

Corollary 3.1. φ_{MR}, φ_{SR}, φ_M and φ_S are compact.

Proof. Recall that independence is preserved under convergence in law. Thus if in the above proof ξ_n^m is independent of X_i^m, $i \leq n$, and ξ_i^m, $i < n$ for m = 1, 2,... with n fixed, it will be so for $m = \infty$. Similarly, if $\{\xi_i^m\}$ are identically distributed for each fixed m = 1,2,..., they remain so for $m = \infty$. The claims for φ_{MR} and φ_{SR} are immediate. Those for φ_M, φ_S follow from the further observation that a limit of Dirac measures in $\mathbb{P}(L)$ is again a Dirac measure. QED

The above proofs in fact lead to the stronger conclusion that the attainable laws of $[(X_0, \xi_0), (X_1, \xi_1),...]$ are compact in $\mathbb{P}((S \times L)^\infty)$, as the CS varies over all CS, MRS, SRS, MS or SS with the law of X_0 being held fixed. One can go a step further and allow the law of X_0 to vary over a compact subset of $\mathbb{P}(S)$. However, the apparently weaker conclusions above will suffice for our purposes.

Corollary 3.2. The law of $[X_0, X_1, X_2,...]$ under an SRS $\gamma[\Phi]$ (resp. an SS $\gamma\{\xi\}$) is a continuous function of Φ (resp. ξ) when viewed as a map from $\mathbb{P}_0(L)$ to $\mathbb{P}(S^\infty)$ (resp. L to $\mathbb{P}(S^\infty)$).

This is immediate in view of the foregoing.

Theorem 3.3. φ_C is convex.

Proof. Let $\{X_n\}$, $\{Y_n\}$ be controlled Markov chains governed by the CS $\{\xi_n\}$, $\{\varphi_n\}$ respectively with $X_0 = Y_0 = 1$. For $n \geq 0$, let $Q_n^1, Q_n^2 \in \mathbb{P}(S^{n+1} \times D^n)$ (with $S \times D^0 = S$ by convention). Denote the laws of $[X_0, X_1,...,X_n, \xi_0(X_0), ..., \xi_{n-1}(X_{n-1})]$ and $[Y_0, Y_1,...,Y_n, \varphi_0(Y_0),..., \varphi_{n-1}(Y_{n-1})]$ respectively ($[X_0]$, $[Y_0]$ resp. when $n = 0$). Let $\alpha_1, \alpha_2 \in [0, 1]$ with $\alpha_1 + \alpha_2 = 1$. Let $Q_n = \alpha_1 Q_n^1 + \alpha_2 Q_n^2$, $n \geq 0$. We shall show that for each n, Q_n is the law of $[Z_0,..,Z_n, \psi_0(Z_0),..., \psi_{n-1}(Z_{n-1})]$ for some controlled Markov chain $\{Z_n\}$ governed by a CS $\{\psi_n\}$. This will imply the statement of the theorem by virtue of the Kolmogorov extension theorem. We proceed by induction. The claim is trivial for $n = 0$. Suppose it is true for some $n \geq 0$. For the sake of simplicity, we let $n \geq 1$, the modifications needed for the case $n = 0$ being quite self-evident. Let $s = [x_0, x_1,...,x_n, y_0,...,y_{n-1}] \in S^{n+1} \times D^n$ and $\bar{s} = [x_0, x_1,...,x_{n+1}, y_0,...,y_n] \in S^{n+2} \times D^{n+1}$ denote typical elements of the respective spaces, to be used as variables of integration in what follows. For $i = 1,2$, let $s \to \eta^i(s,.) : S^{n+1} \times D^n \to \mathbb{P}(D)$ denote any one representative of the regular conditional law of $\xi_n(X_n)$ (resp. $\varphi_n(Y_n)$) given $[X_0, X_1,...,X_n, \xi_0(X_0),...,\xi_{n-1}(X_{n-1})]$ (resp., $[Y_0,...,Y_n, \varphi_0(Y_0),...,\varphi_{n-1}(Y_{n-1})]$), defined a.s. uniquely with respect to the law of the latter. Let $B \subset S^{n+1}$, $B' \subset S$, $C \subset D^n$, $C' \subset D$ be measurable sets. Then

$$Q_{n+1}(B \times B' \times C \times C') = \sum_{i=1}^{2} \alpha_i Q_{n+1}^i (B \times B' \times C \times C')$$

$$= \sum_{i=1}^{2} \alpha_i \sum_{x_{n+1} \in B'} \int_{B \times C} Q_n^i (ds) \int_{C'} p(x_n, x_{n+1}, y_n)\eta^i(s, dy_n).$$

Define a measure $\bar{Q} \in M(S^{n+1} \times D^{n+1})$ by

$$\bar{Q}(B \times C \times C') = \sum_{i=1}^{2} \alpha_i \int_{B \times C} Q_n^i (ds) \int_{C'} \eta^i(s, dy_n).$$

The image of $\bar{Q}$ under the projection $S^{n+1} \times D^{n+1} \to S^{n+1} \times D^n$ is clearly given by

$$\sum_{i=1}^{2} \alpha_i Q_n^i = Q_n.$$

Thus $\bar{Q}$ can be disintegrated as

$$\bar{Q}(ds, dy_n) = Q_n(ds)\, \eta(s, dy_n)$$

where $s \to \eta(s,.) : S^{n+1} \times D^n \to \mathbb{P}(D)$ is any representative of the regular conditional law, defined Q_n - a.s. By the induction hypothesis, Q_n is the law of $[Z_0,\ldots,Z_n, \psi_0(Z_0),\ldots, \psi_{n-1}(Z_{n-1})]$ for some controlled Markov chain Z_i, $0 \le i \le n$, governed by a CS ψ_i, $0 \le i < n$. By enlarging the underlying probability space of these processes if necessary (e.g. by attaching to it a copy of D), construct on it a D-valued random variable $\psi_n(Z_n)$ such that the regular condtional law of $\psi_n(Z_n)$ given $[Z_0,\ldots,Z_n, \psi_0(Z_0),\ldots,\psi_{n-1}(Z_{n-1})]$ is $\eta([Z_0,\ldots,Z_n, \psi_0(Z_0),\ldots,\psi_{n-1}(Z_{n-1})],.)$. By a further enlargement of this probability space if necessary (e.g. by attaching to it a copy of S), construct on it an S-valued random variable Z_{n+1} such that the regular conditional law of Z_{n+1} given $[Z_0,\ldots,Z_n, \psi_0(Z_0),\ldots,\psi_n(Z_n)]$ is $p(Z_n,.,\psi_n(Z_n))$. By construction, the law of $[Z_0,\ldots,Z_{n+1}, \psi_0(Z_0),\ldots, \psi_n(Z_n)]$ is Q_{n+1}. For $i \neq Z_n$, set $\psi_n(i)$ = an arbitrary element of D. By construction, Z_m, $0 \le m \le n+1$, is a controlled Markov chain governed by ψ_m, $0 \le m \le n$. This completes the induction step, proving the claim.

Alternative Proof: Let $Q^1, Q^2 \in \mathbb{P}((S \times D)^\infty)$ denote the respective laws of $[(X_0, \xi_0(X_0)), (X_1, \xi_1(X_1)),\ldots]$ and $[(Y_0, \varphi_0(Y_0)), (Y_1, \varphi_1(Y_1)),\ldots]$ where $\{X_n\}$ (resp. $\{Y_n\}$) is a controlled Markov chain governed by the CS $\{\xi_n\}$ (resp. $\{\varphi_n\}$). Let $\alpha \in [0, 1]$. Define a probability space $(\Omega, \mathcal{F}, P)$ as follows: Let $\Omega = [0, 1] \times (S \times D)^\infty \times (S \times D)^\infty$ and $\mathcal{F}$ be its Borel σ-field. Let P be the product measure: (Lebesgue measure) $\times\, Q^1 \times Q^2$. Denote a typical element of Ω as $w = [a, (x_{1i}, y_{1i}, i \ge 0), (x_{2i}, y_{2i}, i \ge 0)]$ where $a \in [0, 1]$, $x_{ji} \in S$, $y_{ji} \in D$ for all i,j. Define an $(S \times L)$-valued stochastic process (Z_n, ψ_n), $n \ge 0$, on $(\Omega, \mathcal{F}, P)$ as follows:

$$Z_n(w) = x_{1n}\, I\{0 \le a \le \alpha\} + x_{2n}\, I\{\alpha < a \le 1\}$$

$$\psi_n(Z_n(w)) = y_{1n}\, I\{0 \le a \le \alpha\} + y_{2n}\, I\{\alpha < a \le 1\}$$

$$\psi_n(j) = \text{an arbitrary fixed element of D for } j \neq Z_n.$$

It is clear from the construction that $\{Z_n\}$ is a controlled Markov chain governed by the CS $\{\psi_n\}$ and the law of $(Z_i, \psi_i(Z_i))$, $i \ge 0$, is $\alpha Q^1 + (1 - \alpha) Q^2$. QED

Chapter III
The Discounted Cost Problem

This chapter studies the infinite horizon discounted cost control problem. The first section, based on [Bor 5] and [Bor 7], establishes the existence of an optimal SS. The second studies the associated dynamic programming equations.

III.1 The Discounted Occupation Measure

Recall the definitions of the discounted occupation measures $\nu_{Di}, \nu_{D\pi}$ from equations (II.2.7), (II.2.8). For $\pi \in \mathbb{P}(S)$, define $A_{CS}(\pi) = \{\nu_{D\pi} | \{X_n\}$ governed by some CS with the law of $X_0 = \pi\}$. Define $A_{SRS}(\pi)$, $A_{MRS}(\pi)$, $A_{SS}(\pi)$, $A_{MS}(\pi)$ accordingly with respectively SRS, MRS, SS or MS replacing CS in the definition. We shall denote by $\bar{\nu}_{D\pi}$ (resp. $\bar{\nu}_{Di}$) the image of $\nu_{D\pi}$ (resp. ν_{Di}) under the projection $S \times D \to S$.

Theorem 1.1. $A_{CS}(\pi) = A_{SRS}(\pi)$.

Proof. Let $\{X_n\}$ be govened by a CS $\{\xi_n\}$ with initial law π and let $\nu_{D\pi}$ be the corresponding discounted occupation measure. Disintegrate it as

$$\nu_{D\pi}(i, du) = \bar{\nu}_{D\pi}(i)\,\hat{\Phi}_i\,(du) \tag{1.1}$$

where $i \to \hat{\Phi}_i(\cdot): S \to \mathbb{P}(D)$ is the 'regular conditional law', defined $\nu_{D\pi}$ a.s. (Consider if necessary $\nu_{D\pi}, \bar{\nu}_{D\pi}$ to be normalized so as to make them probability measures.) Pick any one representative of this a.s. equivalence class and keep it fixed henceforth. Let $\Phi = \prod_i \hat{\Phi}_i$ and consider a chain $\{X'_n\}$ governed by $\gamma[\Phi]$ with initial law π. Write $\{\xi'_n\} \sim \gamma[\Phi]$. Pick $f \in C_b(S \times D)$ and define $g : S \to R$ by

$$g(i) = E\,[\sum_n \beta^n f(X'_n, \xi'_n(X'_n))/X'_0 = i],\ i \in S.$$

Then g is bounded and satisfies

$$g(i) = \int f(i,u)\hat{\Phi}_i(du) + \beta \sum_{j \in S} g(j)\, p_\Phi(i,j),\ i \in S. \tag{1.2}$$

Define $Z_0 = g(X_0)$ and

$$Z_n = \sum_{m=0}^{n-1} \beta^m f(X_m, \xi_m(X_m)) + \beta^n g(X_n), \;\; n \geq 1,$$

$$W_n = Z_{n+1} - Z_n$$

$$= \beta^n f(X_n, \xi_n(X_n)) + \beta^{n+1} g(X_{n+1}) - \beta^n g(X_n), \;\; n \geq 0.$$

Then

$$E\Big[\sum_{m=0}^{n} \beta^m f(X_m, \xi_m(X_m))\Big] - E\,[g(X_0)\,]$$

$$= E\Big[\sum_{m=0}^{n} W_m\Big] - \beta^{n+1} E\,[\,g(X_{n+1})]. \tag{1.3}$$

The sequence

$$\sum_{m=0}^{n} (W_m - E\,[W_m/\mathcal{F}_m]\,)$$

is a zero mean $\{\mathcal{F}_{n+1}\}$-martingale with bounded increments. Thus

$$E\Big[\sum_{m=0}^{n} W_m\Big] = E\Big[\sum_{m=0}^{n} E\,[W_m/\mathcal{F}_m]\Big]$$

$$= E\Big[\sum_{m=0}^{n} \beta^m \Big(f(X_m, \xi_m(X_m)) + \beta \sum_{j \in S} p(X_m, j, \xi_m(X_m))\;\; g(j) - g(X_m)\Big)\Big]. \tag{1.4}$$

Substitute (1.4) in (1.3) and let $n \to \infty$. By the dominated convergence theorem, we have

$$E\Big[\sum_{m=0}^{\infty} \beta^m f(X_m, \xi_m(X_m))\Big] - E\,[g(X_0)\,]$$

$$= E\Big[\sum_{m=0}^{\infty} \beta^m \Big(f(X_m, \xi_m(X_m)) + \beta \sum_{j \in S} p(X_m, j, \xi_m(X_m))\, g(j) - g(X_m)\Big)\Big]$$

$$= E\Big[\sum_{m=0}^{\infty} \beta^m \Big(\int f(X_m, u)\, \hat{\Phi}_{X_m}(du) + \beta \sum_{j \in S} p_{\Phi}(X_m, j)\, g(j) - g(X_m)\Big)\Big]$$

... (by our construction of Φ)

$= 0$ ... (by virtue of (1.2)).

Thus

$$E\Big[\sum_{m=0}^{\infty} \beta^m f(X_m, \xi_m(X_m))\Big] = E(g(X_0))$$

$$= E[g(X'_0)]$$

$$= E\Big[\sum_{m=0}^{\infty} \beta^m f(X'_m, \xi'_m(X'_m))\Big].$$

Since $f \in C_b(S \times D)$ was arbitrary, the claim follows. QED

This result allows us to restrict our attention to SRS a priori.

Let ν_D denote the infinite matrix whose i'th row is ν_{Di}, $i \in S$. Let I denote the infinite identity matrix. Let M_D denote the set of infinite non-negative matrices whose rows are elements of $M(S)$ with a common bound on their total mass. That is, for $[[a_{ij}]] \in M_D$, $[a_{i1}, a_{i2}, ...]$ is of the form $[\mu(\{1\}), \mu(\{2\}), ...]$ (or simply $[\mu(1), \mu(2), ...]$) for some $\mu \in M(D)$ and $\sup_i \sum_j a_{ij} < \infty$. The following lemma characterizes ν_D in M_D.

Lemma 1.1. Let $\gamma[\Phi]$ be an SRS and ν_D the corresponding matrix defined as above. Then ν_D is the unique solution in M_D satisfying either of the following systems of equations:

$$\nu_D - \beta\, \nu_D P[\Phi] = I, \tag{1.5}$$

$$\nu_D - \beta\, P[\Phi] \nu_D = I. \tag{1.6}$$

Proof. From the definition of ν_{Di}, $i \in S$, we have

$$\nu_{Di}(j) = E\Big[\sum_{n=0}^{\infty} \beta^n I\{X_n = j\}/X_0 = i\Big]$$

where $\{X_n\}$ is governed by $\gamma[\Phi]$. Thus

$$\nu_{Di}(j) = \delta_{ij} + \beta E\Big[E \sum_{n=1}^{\infty} \beta^{n-1} I\{X_n = j\}/X_1\Big]/X_0 = i\Big]$$

$$= \delta_{ij} + \beta \sum_{k \in S} \nu_{Dk}(j)\, p_\Phi(i, k),$$

which yields (1.6). Iterating (1.6),

$$\nu_D = I + \beta P[\Phi] \nu_D = \sum_{n=0}^{m} \beta^n (P[\Phi])^n + \beta^{m+1}(P[\Phi])^{m+1} \nu_D,$$

for $m \geq 1$. Letting $m \to \infty$.

$$\nu_D = \sum_{n=0}^{\infty} \beta^n (P[\Phi])^n, \tag{1.7}$$

proving the uniqueness of ν_D as a solution to (1.6) in M_D. From (1.7), one easily checks that ν_D satisfies (1.5). Argue as above iterating (1.5) instead of (1.6) to conclude that ν_D given by (1.7) is the unique solution to (1.5) in M_D. QED

Let e_i, $i \in S$, denote the i'th row of I and e_π, $\pi \in \mathbb{P}(S)$ the row vector $\sum_i \pi(i)e_i$.

Corollary 1.1. Under an SRS $\gamma[\Phi]$, ∇_{Di}, $\nabla_{D\pi}$ are the unique solutions in M(S) to

$$\nabla_{Di} = \beta \nabla_{Di} P[\Phi] + e_i \tag{1.8}$$

$$\nabla_{D\pi} = \beta \nabla_{D\pi} P[\Phi] + e_\pi \tag{1.9}$$

respectively.

Proof. (1.8) is already contained in (1.5) and (1.9) is immediate form (1.8). Uniqueness follows by iterating the respective equations as in the above lemma. QED

Theorem 1.2. The set $A_{SRS}(\pi) = A_{CS}(\pi)$ is compact convex.

Proof. Compactness follows from Corollary II.3.1 and convexity from Theorem II.3.3. We give an alternative proof of convexity here which will be useful elsewhere as well. Let $\gamma[\Phi_1], \gamma[\Phi_2]$ be two SRS with $\Phi_i = \prod_j \hat{\Phi}_{ij}$ for $i = 1,2$. Fix $\pi \in M(S)$ and let $\nu_{D\pi}[\varphi]$ denote the $\nu_{D\pi}$ under an SRS $\gamma[\varphi]$. Let $a \in [0, 1]$. Define $\Phi = \prod_j \hat{\Phi}_i \in \mathbb{P}_0(L)$ by

$$\hat{\Phi}_i = (a \nabla_{D\pi}[\Phi_1](i)\, \hat{\Phi}_{1i} + (1 - a)\nabla_{D\pi}[\Phi_2](i)\, \hat{\Phi}_{2i}) \,/$$

$$(a \nabla_{D\pi}[\Phi_1](i) + (1 - a) \nabla_{D\pi}[\Phi_2](i)) \tag{1.10}$$

for $i \in$ support $(\nabla_{D\pi}[\Phi_1]) \cup$ support $(\nabla_{D\pi}[\Phi_2])$, and arbitrary otherwise. Using (1.9) for $\nabla_{D\pi}[\Phi_j]$, $j = 1,2$, one obtains

$$(a\,\nabla_{D\pi}[\Phi_1] + (1-a)\,\nabla_{D\pi}[\Phi_2]\,) = e_\pi + \beta(a\nabla_{D\pi}[\Phi_1] + (1-a)\nabla_{D\pi}[\Phi_2]\,)\,P[\Phi].$$

By Corollary 1.1 above,

$$a\,\nabla_{D\pi}[\Phi_1] + (1-a)\,\nabla_{D\pi}[\Phi_2] = \nabla_{D\pi}[\Phi].$$

In view of (1.10), we have

$$a\,\nabla_{D\pi}[\Phi_1]\,(i)\,\hat{\Phi}_{1i} + (1-a)\,\nabla_{D\pi}[\Phi_2]\,(i)\,\hat{\Phi}_{2i} = \nabla_{D\pi}[\Phi]\,(i)\,\hat{\Phi}_i,\ \ i \in S,$$

that is,

$$a\,\nu_{D\pi}[\Phi_1] + (1-a)\,\nu_{D\pi}[\Phi_2] = \nu_{D\pi}[\Phi].$$

This completes the proof. QED

Remark. An analogous argument proves the convexity of

$$\bigcup_{\pi \in \mathbb{P}(S)} A_{SRS}(\pi) = \bigcup_{\pi \in \mathbb{P}(S)} A_{CS}(\pi).$$

Theorem 1.3. $A_{SS}(\pi)$ is a compact subset of $A_{SRS}(\pi)$ and contains all extreme points of the latter.

Proof. Compactness follows from Corollary II.3.1. Let $\gamma[\Phi]$, $\Phi = \prod_i \hat{\Phi}_i$, be an SRS such that there exist some $i_0 \in S$, $a \in (0, 1)$ and $\varphi_1, \varphi_2 \in \mathbb{P}(D)$ such that

$$\nabla_{D\pi}[\Phi]\,(i_0) > 0, \tag{1.11}$$

$$\int p(i_0,.,u)\,\hat{\Phi}_{i_0}(du) = a\int p(i_0,.,u)\varphi_1(du) + (1-a)\int p(i_0,.,u)\varphi_2(du), \tag{1.12}$$

$$\int p(i_0,.,u)\varphi_1(du) \neq \int p(i_0,.,u)\,\varphi_2(du), \tag{1.13}$$

Here (1.11) and (1.12) are to be viewed as vector relations, the integrations being termwise. By relabelling S if necessary, set $i_0 = 1$. Define $\Phi_i \in \mathbb{P}_0(L)$, $i = 1,2$, by $\Phi_i = \varphi_i \times \prod_{j=2}^{\infty} \hat{\Phi}_j$. By (1.11), $\{X_n\}$ governed by $\gamma[\Phi]$ with initial law π hits state 1 with strictly positive probability. Since $P[\Phi]$, $P[\Phi_i]$, $i = 1,2$, differ only in the first row, i.e. in transition probabilities for leaving state 1, the above remark and hence (1.11) also holds with Φ_i, $i = 1,2$, replacing Φ. If any two of $\nabla_{D\pi}[\Phi]$, $\nabla_{D\pi}[\Phi_1]$, $\nabla_{D\pi}[\Phi_2]$ are identical, a simultaneous consideration of the corresponding statements of (1.9) is seen to contradict (1.13) in view of the above observation. Thus any two of them must be distinct. Let $b \in (0,1)$ be such that

$$a = b\,\nabla_{D\pi}[\Phi_1](1)/(b\nabla_{D\pi}[\Phi_1](1) + (1 - b)\nabla_{D\pi}[\Phi_2](1)).$$

We argue as in the proof of Theorem 1.2 to conclude that

$$\nabla_{D\pi}[\Phi] = b\,\nabla_{D\pi}[\Phi_1] + (1 - b)\nabla_{D\pi}[\Phi_2]$$

and thus

$$\nu_{D\pi}[\Phi] = b\,\nu_{D\pi}[\Phi_1] + (1 - b)\,\nu_{D\pi}[\Phi_2].$$

Therefore $\nu_{D\pi}[\Phi]$ cannot be an extreme point of $A_{SRS}(\pi)$. This implies that if $\nu_{D\pi}[\Phi]$ is an extreme point of $A_{SRS}(\pi)$ for some $\gamma[\Phi]$, $\Phi = \prod_i \hat{\Phi}_i$, the following must hold: For each i in the support of $\nabla_{D\pi}[\Phi]$ (denoted $K[\Phi]$ henceforth), $p(i,.,u)$ is the same for all $u \in$ support $(\hat{\Phi}_i)$. Hence $P[\Phi'] = P[\Phi]$ for all SRS $\gamma[\Phi']$, $\Phi' = \pi\hat{\Phi}'_i$, such that support $(\hat{\Phi}'_i)$ $\subset$ support $(\hat{\Phi}_i)$ for $i \in K[\Phi]$, and $\hat{\Phi}_i = \hat{\Phi}'_i$ otherwise. Thus $\nabla_{D\pi}[\Phi] = \nabla_{D\pi}[\Phi']$ for such $\gamma[\Phi']$. Suppose that for some $i \in K[\Phi]$, say $i = 1$, $\hat{\Phi}_i = \hat{\Phi}_1 = \alpha\varphi_1 + (1 - \alpha)\varphi_2$ for some $\alpha \in (0, 1)$ and $\varphi_1 \neq \varphi_2$ in $\mathbb{P}(D)$. Set $\Phi_1 = \varphi_1 \times \prod_{i=2}^{\infty} \hat{\Phi}_i$, $\Phi_2 = \varphi_2 \times \prod_{i=2}^{\infty} \hat{\Phi}_i$. Then by the foregoing, $\nabla_{D\pi}[\Phi] = \nabla_{D\pi}[\Phi_i]$, $i = 1,2$. It is easy to check that

$$\nu_{D\pi}[\Phi] = \alpha\nu_{D\pi}[\Phi_1] + (1 - \alpha)\nu_{D\pi}[\Phi_2], \quad \nu_{D\pi}[\Phi_1] \neq \nabla_{D\pi}[\Phi_2],$$

contradicting the fact that $\nu_{D\pi}[\Phi]$ is an extreme point of $A_{SRS}(\pi)$. Hence $\hat{\Phi}_1$ must be a Dirac measure. Thus $\hat{\Phi}_i$ are Dirac measures for $i \in K[\Phi]$. Since for $i \notin K[\Phi]$ the corresponding rows of $P[\Phi]$ do not enter the picture at all, we may assume that $\hat{\Phi}_i$ are Dirac measures for all $i \in S$, that is, $\nu_{D\pi}[\Phi] \in A_{SS}(\pi)$. QED

In fact, the above proof establishes the following slightly stronger result.

Corollary 1.2. A necessary condition for an element of $A_{SRS}(\pi)$ to be its extreme point is that it be the discounted occupation measure for an SS $\gamma\{\xi\}$, $\xi = [\xi(1), \xi(2),...]$, satisfying the following for each $i \in S$: $p(i,.,\xi(i)) \in \mathbb{P}(S)$ is an extreme point of the set $\{p_\Phi(i,.), \Phi \in \mathbb{P}_0(L)\} \subset \mathbb{P}(S)$.

These results will now be used to deduce the existence of an optimal SS.

Theorem 1.4(a). For any initial law π, there exists an optimal SS $\gamma\{\xi\}$.

Proof. The map $\nu \in A_{CS}(\pi) = A_{SRS}(\pi) \to \int k d\nu \in R$ is lower semicontinuous and hence attains a minimum on $A_{SRS}(\pi)$ in view of its compactness. Suppose the minimum is attained at $\bar{\nu}$. By Choquet's theorem [Phe], $\bar{\nu}$ is the barycenter of a probability measure μ supported on the set of extreme points of $A_{SRS}(\pi)$ and hence, by Theorem 1.3, on $A_{SS}(\pi)$. Then

$$\int k d\bar{\nu} = \int_{A_{ss}(\pi)} \mu(d\nu) \int k d\nu .$$

If $\int k d\nu \neq \int k d\bar{\nu}$ for all $\nu \in A_{SS}(\pi)$, the above would imply that $\int k d\nu < \int k d\bar{\nu}$ for some $\nu \in A_{SS}(\pi)$. This contradicts the fact that $\nu \to \int k d\nu$ attains its minimum on $A_{SRS}(\pi)$ at $\bar{\nu}$. Thus there must exist a $\nu \in A_{SS}(\pi)$ for which $\int k d\nu = \int k d\bar{\nu}$, and the proof is complete.

QED

Once again, a little thought shows that the ξ above can be chosen so as to satisfy the 'necessary condition' in Corollary 1.2.

Lemma 1.2. For any $i \in$ support (π), $\gamma\{\xi\}$ above is optimal for initial condition i.

Proof. Supose it is not. Then there exists an $i \in$ support (π) and an SS $\gamma\{\xi'\}$ such that $\gamma\{\xi'\}$ yields a strictly lower cost than $\gamma\{\xi\}$ when $X_0 = i$. It is easy to see that the CS $\{\bar{\xi}_n\}$ given by

$$\bar{\xi}_n = \xi \, I\,\{X_0 \neq i\} + \xi' \, I\{X_0 = i\}, \;\; n \geq 0,$$

gives a strictly lower cost than $\gamma\{\xi\}$ for initial law π, a contradiction. QED

Theorem 1.4(b) The $\gamma\{\xi\}$ above can be taken to be optimal for any initial law.

Proof. Since $\int k d\nu_{D\pi} = \Sigma\, \pi(i) \int k d\nu_{Di}$, it suffices to prove the above for initial laws which are Dirac measures at some $i \in S$. Consider an initial law π with support $(\pi) = S$ and apply the above lemma. QED

III.2 The Dynamic Programming Equations

Let $V : S \to R^+$ be defined by $V(i) = \int k d\nu_{Di}$ under the optimal SS $\gamma\{\xi\}$ of Theorem 1.4 (a), (b). This section studies the 'dynamic programming equations' (equations (2.1) below) satisfied by V. These lead to a characterization of an optimal SS. Clearly, $V(i) = \min_{CS} \int k d\nu_{Di}$.

For $\varphi \in L$ and $\Phi = \prod_i \hat{\Phi}_i \in \mathbb{P}_0(L)$, define the column vectors $k\{\varphi\} = [k(1, \varphi(1)), k(2, \varphi(2)), \ldots]^T$, $k[\Phi] = [\int k(1, u)\, \hat{\Phi}_i(du), \int k(2,u)\, \hat{\Phi}_2(du), \ldots]^T$. V itself will be identified with the column vector $[V(1), V(2),...]^T$. Since $V(.) \geq 0$, the quantity $\sum_j p_\Phi(i, j) V(j)$ is a well-defined number in $[0, \infty]$.

Theorem 2.1. V satisfies

$$V = \min_\varphi (\beta P\{\varphi\} V + k\{\varphi\}) \tag{2.1}$$

or equivalently,

$$V = \min_\Phi \{\beta P[\Phi] V + k[\Phi]) \tag{2.1'}$$

where the minimum is termwise. Any other solution $V' : S \to R^+$ of (2.1) or (2.1′) must satisfy $V' \geq V$ termwise.

Proof. Let $\gamma\{\xi\}$ be as in Theorem 1.4(a), (b). As in (1.2), we have

$$V = \beta P\{\xi\} V + k\{\xi\}. \tag{2.2}$$

Here each component of $P\{\xi\}V$ is finite (because that of V is). Suppose (2.1) is false. Then for some $i_0 \in S$, $u \in D$ and $\Delta > 0$, we have

$$\sum_{j \in S} p(i_0, j, u) V(j) < \infty,$$

$$V(i_0) = \beta \sum_{j \in S} p(i_0, j, u) V(j) + k(i_0, u) + \Delta. \tag{2.3}$$

Let $\{X_n\}$ be governed by $\gamma\{\xi'\}$ defined by $\xi'(j) = \xi(j)$ for $j \neq i_0$ and $= u$ for $j = i_0$, with $X_0 = i_0$. By (2.2) and (2.3),

$$V(i) = \beta \sum_{j \in S} p(i, j, \xi'(i)) V(j) + k(i, \xi'(i)) + \Delta I\{i = i_0\}.$$

Iterating, this leads to

$$V(i_0) = E\left[\sum_{m=0}^{n} \beta^m k(X_m, \xi'(X_m))\right] + \Delta E\left[\sum_{m=0}^{n} \beta^m I\{X_m = i_0\}\right]$$

$$+ \beta^{n+1} E[V(X_{n+1})]$$

$$> E\Big[\sum_{m=0}^{n} \beta^m k(X_m, \xi'(X_m))\Big] + \Delta E\Big[\sum_{m=0}^{n} \beta^m I\{X_m = i_0\}\Big].$$

Letting $n \to \infty$ on the right hand side, we get

$$V(i_0) \geq \int k d\nu_{Di_0} + \Delta\, \nu_{Di_0}(\{i_0\})$$

$$\geq \int k d\nu_{Di_0} + \Delta$$

where ν_{Di_0} is the discounted occupation measure under $\gamma\{\xi'\}$. This contradicts the optimality of $\gamma\{\xi\}$, establishing (2.1). (2.1') follows similarly. If $V' : S \to R^+$ satisfies (2.1), it follows that the set $A_{i\varepsilon} = \{u \in D \mid \sum_j p(i, j, u)\, V'(j) \leq V'(i) + \varepsilon\}$ is non-empty for $i \in S$, $\varepsilon > 0$. Using Fatou's lemma, one checks that it is also closed and hence compact. Since for each i, $\sum_j p(i,j,u)\, V'(j)$ is an increasing limit of non-negative continuous functions of u, it is lower semicontinuous in u on the above compact non-empty sets. Thus it attains for each i a minimum on $A_{i\varepsilon}$, $\varepsilon > 0$ arbitrary, and hence on D. Pick $\bar{\xi} \in L$ such that

$$P\{\bar{\xi}\}V' + k\{\bar{\xi}\} = \min_{\varphi} (P\{\varphi\}V' + k\{\varphi\}) = V' \tag{2.4}$$

Let ν_{Di}, $i \in S$, denote the discounted occupation measures under $\gamma\{\bar{\xi}\}$. Iterating (2.4), an argument similar to the one used above leads to

$$V'(i) \geq \int k d\nu_{Di} \geq V(i), \quad i \in S.$$

The corresponding claim for $V' : S \to R^+$ satisfying (2.1') follows similarly. QED

Corollary 2.1. If k is bounded, V is the unique solution to (2.1) in $\{f: S \to R^+ \mid \sup_i |f(i)| < \infty\}$.

Proof. If k is bounded, V clearly is. Let V' be another solution of (2.1) in the desired class. Pick $\gamma\{\bar{\xi}\}$ as in (2.4). Let $i \in S$, and $\{X_n\}$ be a chain governed by $\gamma\{\bar{\xi}\}$ with $X_0 = i$. Argue as in the proof of the above lemma to obtain:

$$V'(i) = E\Big[\sum_{m=0}^{n} \beta^m k(X_m, \bar{\xi}(X_m))\Big] + \beta^{n+1} E[V'(X_{n+1})]. \tag{2.5}$$

Letting $n \to \infty$, $V'(i) = \int k d\nu_{Di}$ under $\gamma\{\bar{\xi}\}$. Under any other SS $\gamma\{\varphi\}$, (2.5) holds with = replaced by $\leq$ for all i in view of (2.1). Letting $n \to \infty$, one then has $V'(i) \leq \int k d\nu_{Di}$ under $\gamma\{\varphi\}$. Thus

$$V'(i) = \min_{SS} \int k d\nu_{Di} = V(i), \ i \in S. \qquad \text{QED}$$

The boundedness condition in the above corollary cannot in general be dropped. Consider for example the (uncontrolled) symmetric random walk on integers with 'cost' $k(.,.) = 0$, $\beta = \frac{1}{2}$. Then (2.1) reduces to

$$V(i) = \frac{1}{4}(V(i+1) + V(i-1)), \ i = 0, \pm 1, \pm 2 \dots .$$

Clearly, $V = [0, 0, \dots]^T$ is the desired solution, but other unbounded non-negative solutions are possible.

The importance of (2.1) is due to the following characterization of an optimal SS.

Theorem 2.2. $\gamma\{\xi\}$ is optimal for any initial law if and only if ξ attains the termwise minimum in (2.1).

Proof. If ξ attains the termwise minimum in (2.1), we have

$$V(i) = \beta \sum_{j \in S} p(i,j,\xi(i))\, V(j) + k(i, \xi(i)), \ i \in S.$$

Let $\{X_n\}$ be governed by $\gamma\{\xi\}$ with $X_0 =$ (say) i. Then familiar arguments yield

$$V(i) \geq E\left[\sum_{m=0}^{n} \beta^m k(X_m, \xi(X_m))\right], \ n \geq 1.$$

Letting $n \to \infty$,

$$V(i) \geq \int k d\nu_{Di} \geq V(i),$$

i.e., $\gamma\{\xi\}$ is optimal for initial condition i. Since i was arbitrary, it is optimal for any initial condition. The converse follows from (2.2). QED

Corollary 2.2. An SRS $\gamma[\Phi]$ is optimal for any initial law if and only if Φ attains the termwise minimum in (2.1').

This is proved along similar lines. We now give another variant of Theorem 2.2. For an SS $\gamma\{\varphi\}$, define $V\{\varphi\} = [V\{\varphi\}(1), V\{\varphi\}(2), \dots]^T$ by $V\{\varphi\}(i) = \int k d\nu_{Di}$ for $i \in S$, the measure ν_{Di} being under $\gamma\{\varphi\}$. As in (2.2), one then has

$$V\{\varphi\} = \beta P\{\varphi\} V\{\varphi\} + k\{\varphi\}.$$

Theorem 2.3. $\gamma\{\xi\}$ is optimal for any initial condition if and only if

$$V\{\xi\} = \beta P\{\xi\}\ V\{\xi\} + k\{\xi\} = \min_{\varphi} (\beta P\{\varphi\} V\{\varphi\} + k\{\varphi\}). \tag{2.6}$$

Proof. Let $V\{\xi\}$ be optimal for any initial condition. Since $V\{\varphi\} \geq V\{\xi\}$ termwise for arbitrary $\gamma\{\varphi\}$, the second equality follows. Conversely, let (2.6) hold. Let $\gamma\{\bar{\xi}\}$ be an optimal SS. Then for $\gamma\{\xi\}$ as above,

$$V\{\bar{\xi}\} \leq V\{\xi\} = \min_{\varphi} (\beta P\{\varphi\} V\{\varphi\} + k\{\varphi\})$$

$$\leq \beta P\{\bar{\xi}\} V(\bar{\xi}\} + k\{\bar{\xi}\} = V\{\bar{\xi}\}$$

termwise, implying $V\{\xi\} = V\{\bar{\xi}\}$. Thus $\gamma\{\xi\}$ is optimal. QED.

Chapter IV
Finite Time Control Problems

This chapter studies the problem of control up to a first exit time and the finite horizon control problem, along the lines of the preceding chapter. The principal references are [Bor 5] and [Bor 7].

IV.1 Preliminaries for the exit problem

Consider the cost (C2) defined by (II.2.2). Recall the set A and the stopping time T introduced there. We shall make the following assumption.

(*) There exists an integer $N \geq 1$ and $\delta > 0$ such that under any CS and for any $i \in A$, there exists a path from i to A^c of length not exceeding N whose probability exceeds δ.

A sufficient condition for this to hold is given by the following lemma.

Lemma 1.1. If S is a single communicating class under all SRS, the (*) holds for all finite $A \subset S$.

Proof. Suppose not. Then there exists a sequence of controlled Markov chains $\{X_n^m\}$, $m = 1,2,\ldots$, governed by CS $\{\xi_n^m\}$, $m = 1,2,\ldots$, respectively, with the initial law supported in A, such that the following holds: If $\tau^m = \min\{n \geq 0 \mid X_n^m \notin A\}$ ($= \infty$ if $X_n^m \in A$ for all n), then

$$P(\tau^m > m) > 1 - \frac{1}{m}, \quad m \geq 1.$$

By dropping to a subsequence if necessary, and invoking Skorohod's theorem as in the proof of Theorem II.3.2, we may assume that these chains are defined on a common probability space, and there exists a controlled Markov chain $\{X_n^\infty\}$ governed by a CS $\{\xi_n^\infty\}$ with $X_0^\infty \in A$ a.s., such that

$$[X_0^m, X_1^m, X_2^m, \ldots, \xi_0^m, \xi_1^m, \ldots] \rightarrow [X_0^\infty, X_1^\infty, \ldots, \xi_0^\infty, \xi_1^\infty, \ldots] \text{ a.s.}$$

Since

$$P(\tau^m > j) = E[\prod_{i=1}^{j} I\{X_n^m \in A\}],\ m,j = 1,2, \dots ,$$

a straightforward limiting argument leads to

$$P(\tau^\infty > m) > 1 - \frac{1}{m},\ m \geq 1.$$

for $\tau^\infty = \min\{n \geq 0 \mid X_n^\infty \notin A\}$. Thus $\tau^\infty = \infty$, a.s. This is possible only if there exists a non-empty subset G of A such that for $i \in G$.

$$\sup_{j \notin G} \inf_u p(i, j, u) = 0.$$

(This is easily checked, for example by replacing A^c by a single absorbing state.) Since $p(i, j, .)$ is continuous and D compact, the above infimum is attained at some point u_i in D. Then the chain starting in G and governed by an SS $\gamma\{\xi\}$ such that $\xi = [\xi(1), \xi(2), \dots]$ and $\xi(i) = u_i$ for $i \in G$, does not leave G at any time, a.s. This contradicts the hypothesis that S is a single communicating class under all SRS. Thus $(*)$ must hold. QED

We show next that $(*)$ implies bounded moments of T, in particular implying that the occupation measure up to T, ν_{Ti} defined by (II.2.10)-(II.2.11), is a well-defined measure with a finite total mass.

Corollary 1.1. For $n \geq 1$, $\sup E[(T)^n] < \infty$, where the supremum is over all CS and all initial laws supported in A.

Proof. Let $\{X_n\}$ be a controlled Markov chain governed by a CS $\{\xi_n\}$ with $X_0 \in A$ a.s. and let T be its first exit time from A. Let N, δ be as in $(*)$. Then

$$P(T > N) < 1 - \delta$$

regardless of the CS and the exact law of X_0. Using this and Theorem II.3.1, one has

$$P(T > nN) = E[\prod_{m=0}^{nN} I\{X_m \in A\}]$$

$$= E[E[\prod_{m=(n-1)N+1}^{nN} I\{X_m \in A\}/\mathcal{F}_{(n-1)N}] \prod_{m=0}^{(n-1)N} I\{X_m \in A\}]$$

$$= E[P(T > nN/\mathcal{F}_{(n-1)N})\, I\{T > (n-1)N\}]$$

$$\leq (1-\delta)P(T > (n-1)N).$$

Iterating,

$$P(T > nN) \leq (1-\delta)^n, \quad n \geq 1.$$

The claim follows quite easily from this. QED.

Let $A'_{CS}(\pi)$, $A'_{SRS}(\pi)$, $A'_{SS}(\pi)$ denote the sets of occupation measures up to T under all CS, all SRS and all SS respectively when the initial law is π.

Lemma 1.2. $A'_{CS}(\pi)$ is compact.

Proof. Take $\pi(1) = 1$ with $1 \in A$ for the sake of simplicity. Recall the set-up of Theorem II.3.2. Let T_m be the first exit time from A for the chain $\{X_n^m, n \geq 0\}$ for $m = 1,2, ..., \infty$, and ν_{T1}^m the corresponding occupation measures up to T_m. Then $T_m \to T_\infty$ a.s. and thus for $f \in C(A \times D)$,

$$\sum_{n=0}^{T_m - 1} f(X_n^m, \xi_n^m(X_n^n)) \to \sum_{n=0}^{T_\infty - 1} f(X_n^\infty, \xi_n^\infty(X_n^\infty)) \text{ a.s.}$$

By Corollary 1.1, these random variables are uniformly integrable and thus one may take expectations in the above to obtain:

$$\int f d\nu_{T1}^m \to \int f d\nu_{T1}^\infty .$$

Since $f \in C(A \times D)$ was arbitrary, $\nu_{T1}^m \to \nu_{T1}^\infty$ in $M(A \times D)$. This proves sequential compactness and hence the compactness of $A'_{CS}(\pi)$. QED

The compactness of $A'_{SRS}(\pi)$ and $A'_{SS}(\pi)$ is also proved along similar lines in view of Corollary II.3.1.

The rest of the developments mimic closely those of the preceding chapter. Let $\bar{\nu}_{Ti})$ (resp. $\bar{\nu}_{Ti}, i \in S$) denote the image of $\nu_{T\pi}$ (resp. ν_{Ti}) under the projection $A \times D \to A$.

Theorem 1.1. $A'_{CS}(\pi) = A'_{SRS}(\pi)$.

Proof. Let $\{X_n\}$ be a chain governed by a CS $\{\xi_n\}$ with initial law π and $\nu_{T\pi}$ the corresponding occupation measure up to the first exit time T. Disintegrate $\nu_{T\pi}$ as

$$\nu_{T\pi}(i, du) = \bar{\nu}_{T\pi}(i)\psi_i(du), \quad i \in A,$$

where $i \in A \to \psi_i \in \mathbb{P}(D)$ is the 'regular conditional law' defined $\bar{\nu}_{T\pi}$ - a.s. (Consider normalized versions of $\nu_{T\pi}, \bar{\nu}_{T\pi}$ if necessary.) Take any one representative of this a.s. equivalence class and define an SRS $\gamma[\Phi]$, $\Phi = \prod_i \hat{\Phi}_i$, by: $\hat{\Phi}_i = \psi_i$ for $i \in A$ and arbitrary otherwise. Let $\{X'_n\}$ be a chain governed by the SRS $\{\xi'_n\} \sim \gamma[\Phi]$ with initial law π and let T' be its first exit time from A. Let $f \in C_b(S \times D)$ and define $g : S \to R$ by

$$g(i) = E\Big[\sum_{m=0}^{T'-1} f(X'_n, \xi'_n(X'_n))/X'_0 = i\Big], \quad i \in A,$$

$$g(i) = 0 \text{ otherwise.}$$

Then for $i \in A$, $g(i)$ satisfies

$$g(i) = \int f(i, u)\hat{\Phi}_i(du) + \sum_{j \in S} p_\Phi(i, j)g(j). \tag{1.1}$$

Define

$$Z_0 = g(X_0)$$

$$Z_n = \sum_{m=0}^{n-1} f(X_m, \xi_m(X_m)) + g(X_n), \quad n \geq 1.$$

$$W_n = Z_{n+1} - Z_n$$

$$= f(X_n, \xi_n(X_n)) + g(X_{n+1}) - g(X_n), \quad n \geq 0.$$

Then $\sum_{m=0}^{n} (W_m - E[W_m/\mathcal{F}_m])$, $n \geq 0$, is an $\{\mathcal{F}_{n+1}\}$-martingale with zero mean and bounded increments. Applying to it the optional sampling theorem, one obtains

$$E\Big[\sum_{m=0}^{T-1} f(X_m, \xi_m(X_m))\Big] - E[g(X_0)]$$

$$= E\Big[\sum_{m=0}^{T-1} W_m\Big]$$

$$= E\Big[\sum_{m=0}^{T-1} E[W_m / \mathcal{F}_m]\Big]$$

$$= E\Big[\sum_{m=0}^{T-1} (f(X_m, \xi_m(X_m)) + \sum_{j \in S} g(j)p(X_m, j, \xi_m(X_m)) - g(X_m))\Big]$$

$$= E\Big[\sum_{m=0}^{T-1} (\int f(X_m, u)\hat{\Phi}_{X_m}(du) + \sum_{j \in S} \int p(X_m, j, u)\hat{\Phi}_{X_m}(du)g(j) - g(X_m))\Big]$$

(by the definition of Φ)

$$= 0$$

by (1.1). Thus

$$E\Big[\sum_{m=0}^{T-1} f(X_m, \xi_m(X_m))\Big] = E\big[g(X_0)\big]$$

$$= E[g(X'_0)]$$

$$= E\Big[\sum_{m=0}^{T'-1} f(X'_m, \xi'_m(X'_m))\Big].$$

The claim follows. QED.

Let q denote the cardinality of A, and I_q the $q \times q$ identity matrix with rows $\bar{e}_i$, $1 \le i \le q$. Let ν_T be the $q \times q$ matrix whose i'th row is $\bar{\nu}_{Ti}$, and M_T the set of $q \times q$ non-negative matrices whose rows are elements of $M(A)$.

Lemma 1.3. Let $\nu[\Phi]$ be an SRS and ν_T the corresponding matrix defined as above. Then ν_T is the unique solution in M_T of either of the following systems of equations:

$$\nu_T - \nu_T \bar{P}[\Phi] = I_q \tag{1.2}$$

$$\nu_T - \bar{P}[\Phi]\, \nu_T = I_q \tag{1.3}$$

where $\bar{P}[\Phi]$ is the submatrix of $P[\Phi]$ obtained by taking the rows and columns corresponding to the elements of A, ordered as in ν_T.

Proof. Let $\{X_n\}$ be a chain governed by $\gamma[\Phi]$. Then for $i, j \in A$,

$$\bar{\nu}_{Ti}(j) = E\Big[\sum_{n=0}^{T-1} I\{X_n = j\}/X_o = i\Big]$$

$$= \delta_{ij} + E\Big[E\big[\sum_{n=1}^{T-1} I\{X_n = j\}/X_1\big] I\{X_1 \in A\}/X_0 = i\Big]$$

$$= \delta_{ij} + \sum_{k \in A} p_{\Phi}(i, k)\bar{\nu}_{Tk}(j),$$

which yields (1.3). Iterating (1.3),

$$\nu_T = I_q + \bar{P}[\Phi]\nu_T = \sum_{n=0}^{m} (\bar{P}[\Phi])^n + (\bar{P}[\Phi])^{m+1}\nu_T, \quad m \geq 1.$$

The (i,j)'th element of $(\bar{P}[\Phi])^{m+1}$ is the probability under $\gamma[\Phi]$ of going from i to j in $m+1$ steps without having exited from A in the meantime. This is bounded from above by $P(T > m+1)$ which tends to zero as $m \to \infty$. Letting $m \to \infty$ in the above, we then have

$$\nu_T = \sum_{n=0}^{\infty} (\bar{P}[\Phi])^n. \tag{1.4}$$

This proves the uniqueness of ν_T as a solution to (1.3) in M_T. From (1.4), one easily checks that ν_T also satisfies (1.2). Iterating (1.2) in place of (1.3), one recovers (1.4), showing that ν_T is the unique solution to (1.2) in M_T. QED.

Let $\bar{e}_\pi = \sum_{i \in A} \pi(i)\bar{e}_i$, $\pi \in \mathbb{P}(A)$.

Corollary 1.2. Under an SRS $\gamma[\Phi]$ as above, $\bar{\nu}_{Ti}$ and $\bar{\nu}_{T\pi}$ are unique solutions in $M(A)$ to

$$\bar{\nu}_{Ti} = \bar{\nu}_{Ti}\bar{P}[\Phi] + \bar{e}_i, \tag{1.5}$$

$$\bar{\nu}_{T\pi} = \bar{\nu}_{T\pi}\bar{P}[\Phi] + \bar{e}_\pi \tag{1.6}$$

respectively.

Proof. (1.5) is contained in the above lemma. (1.6) is immediate from (1.5). Uniqueness follows by iterating the respective equations and arguing as in the above lemma. QED.

Theorem 1.2. The set $A'_{SRS}(\pi) = A'_{SS}(\pi)$ is compact convex.

Proof. Compactness is proved in Lemma 1.2. Convexity follows from Theorem II.3.3, or alternatively along the lines of Theorem III.1.2 with (1.6) playing the role of (III.1.9). QED.

A remark similar to the one following Theorem III.1.2 applies here as well.

Theorem 1.3. $A'_{SS}(\pi)$ is a compact subset of $A'_{SRS}(\pi)$ and contains all extreme points of the latter.

Proof. As already observed, compactness follows as in Lemma 1.2 in view of Corollary II.3.1. The rest follows along the lines of Theorem III.1.3 with (1.6) playing the role of (III.1.9). QED.

The obvious analogue of Corollary III.1.2 holds here for obvious reasons.

IV.2. Existence and Dynamic Programming

Returning to the control problem (C2), we now have the following.

Theorem 2.1. An optimal SS $\gamma\{\xi\}$, optimal for any initial law $\pi \in \mathbb{P}(A)$, exists.

Given the results of the preceding section, the proof of this theorem is virtually identical to that of Theorem 1.4 (a), (b), and is omitted.

Let $\partial A = \{j \in S\backslash A \mid p(i, j, u) > 0$ for some $i \in A$ and $u \in D\}$ and $\bar{A} = A \cup \partial A$. Define $V : \bar{A} \to R^+$ by

$$V(i) \begin{cases} = h(i) & \text{for } i \in \partial A, \\ = \int \bar{k} d\nu_{Ti} - h(i) & \text{for } i \in A, \end{cases}$$

where ν_{Ti} is the occupation measure up to T under the optimal SS $\gamma\{\xi\}$ of the above theorem, and $\bar{k}$ is defined as in (II.2.12). Clearly,

$$\begin{aligned} V(i) &= \min_{CS} \left(\int \bar{k} d\nu_{Ti} - h(i)\right) \\ &= \min_{CS} E\left[\sum_{m=0}^{T-1} f(X_m, \xi_m(X_m)) + h(X_T)/X_0 = i\right] \quad (2.1) \end{aligned}$$

for $i \in A$. The 'dynamic programming equations' for (C2) are equations (2.2) and (2.3) below, satisfied by the 'value function' V defined above.

Theorem 2.2. V is the unique solution in $C(\bar{A})$ to

$$V(i) = \min_u \Big(\sum_{j \in S} p(i, j, u)V(j) + k(i, u) \Big), \; i \in A \tag{2.2}$$

$$V(i) = h(i), \; i \in \partial A, \tag{2.3}$$

or equivalently,

$$V(i) = \min_\Phi \Big(\sum_{j \in S} V(j) \int p(i, j, u)\Phi(du) + \int k(i, u)\Phi(du) \Big), \; i \in A, \tag{2.2'}$$

along with (2.3).

Proof. Suppose (2.2) is false. Then for some $i_0 \in A$, $\Delta > 0$ and $u \in D$,

$$V(i_0) = \sum_{j \in S} p(i_0, j, u)V(j) + k(i_0, u) + \Delta. \tag{2.4}$$

From the definition of V, it follows that

$$V(i) = \sum_i p(i, j, \xi(i))V(j) + k(i, \xi(i)), \; i \in A. \tag{2.5}$$

(Compare with (1.1).) Define an SS $\gamma\{\xi'\}$ by: $\xi'(i) = \xi(i)$ for $i \neq i_0$ and $= u$ for $i = i_0$. Let $\{X_n\}$ be governed by $\gamma\{\xi'\}$ with $X_0 = i_0$. Then for T as before, $\{M_n\}$ defined by $M_0 = 0$,

$$M_{n+1} = \sum_{m=0}^{n \wedge T} (V(X_{m+1}) - \sum_{j \in S} p(X_m, j, \xi'(X_m))V(j)), \; n \geq 0,$$

is a zero mean bounded martingale. By the optional sampling theorem, $E[M_T] = 0$. In view of (2.3), (2.4) and (2.5), this is equivalent to

$$0 = E\Big[\sum_{m=0}^{T-1} (V(X_{m+1}) - \sum_{j \in S} p(X_m, j, \xi'(X_m))V(j)) \Big]$$

$$= E[h(X_T)] + E\Big[\sum_{m=0}^{T-1} (V(X_m) - \sum_{j \in S} p(X_m, j, \xi'(X_m))V(j))\Big] - V(i_0)$$

$$= E[h(X_T) + \sum_{m=0}^{T-1} k(X_m, \xi'(X_m))] + \Delta E\Big[\sum_{m=0}^{T-1} I\{X_m = i_0\} \Big] - V(i_0).$$

Since $X_0 = i_0$,

$$V(i_0) > E[h(X_T) + \sum_{m=0}^{T-1} k(X_m, \xi'(X_m))],$$

a contradiction to the optimality of $\gamma\{\xi\}$. Thus (2.2) must hold. If V' is another solution of (2.2) and (2.3), pick $\bar{\xi} \in L$ such that

$$\sum_{j \in S} p(i, j, \bar{\xi}(i))V'(j) + k(i, \bar{\xi}(i)) = \min_{\varphi} (\sum_{j \in S} p(i, j, \varphi(i))V'(j) + k(i, \varphi(i))),$$

where the minimum exists by arguments similar to those for (III.2.4). Let $\{X_n\}$ be governed by $\gamma\{\bar{\xi}\}$ with $X_0 = i$. An argument analogous to the above leads to

$$V'(i) \geq E\Big[\sum_{m=0}^{T-1} k(X_m, \bar{\xi}\{X_m)) + h(X_T) \Big] \geq V(i).$$

On the other hand,

$$V'(i) \leq \sum_{j \in S} p(i, j, \xi(i))V'(j) + k(i, \xi(i)), \ \ i \in A.$$

Consider $\{X_n\}$ governed by $\gamma\{\xi\}$ with $X_0 = i$, and argue as above to conclude

$$V'(i) \leq E\Big[\sum_{m=0}^{T-1} k(X_m, \xi(X_m)) + h(X_T) \Big] = V(i).$$

Thus $V' = V$. The parallel claim for (2.2′) and (2.3) follows along similar lines. QED.

Theorem 2.3. $\gamma\{\xi\}$ is optimal for any initial law if and only if ξ attains the termwise minimum in (2.2).

Proof. If ξ attains the termwise minimum in (2.2), we have

$$V(i) = \sum_{j \in S} p(i, j, \xi(i))V(j) + k(i, \xi(i)), \ \ i \in S.$$

Let $\{X_n\}$ be governed by $\gamma\{\xi\}$ with $X_0 = i$. Then one can argue as above to conclude that

$$V(i) = E\Big[\sum_{m=0}^{T-1} k(X_m, \xi(X_m)) + h(X_T) \Big].$$

In view of (2.1) and the arbitrariness of i, it follows that $\gamma\{\xi\}$ is optimal for any initial condition. Conversely, let $\gamma\{\xi\}$ be as in Theorem 2.1. Then ξ attains the termwise minimum in (2.2) by virtue of (2.5). QED.

The obvious analogue of Corollary III.2.2, i.e. the above with $\gamma[\Phi]$, (2.2′) replacing $\gamma\{\xi\}$, (2.2), also holds. Finally, define for each SS $\gamma\{\xi\}$ the function $V\{\xi\} : \bar{A} \to R^+$ by

$$V\{\xi\}(i) = E\Big[\sum_{m=0}^{T-1} k(X_m, \xi_m(X_m)) + h(X_T)\Big] \geq V(i), \ \ i \in A,$$

$$V\{\xi\}(i) = h(i), \ \ i \in \partial A,$$

where $\{X_n\}$ is governed by $\gamma\{\xi\}$ with $X_0 = i$. We then have the following analogue of Theorem III.2.3.

Theorem 2.4. $\gamma\{\xi\}$ is optimal for any initial condition if and only if

$$V\{\xi\}(i) = \sum_{j \in S} p(i, j, \xi(i))V\{\xi\}(j) + k(i, \xi(i))$$

$$= \min_{\varphi} \Big(\sum_{j \in S} p(i, j, \varphi(i))V\{\varphi\}(j) + k(i, \varphi(i))\Big)$$

for $i \in A$.

The proof is virtually identical to that of Theorem III.2.3, with obvious modifications such as replacing (III.2.1) by (2.2) above.

We conclude our study of control up to a first exit time with one related remark. Let $\{X_n\}$, $\{X'_n\}$ be as in the proof of Theorem 1.1. Define $\bar{k}$ as before but with $k(.,.) \equiv 0$. Arguing as in the proof of Theorem 1.1, one then has $E[h(X_T)] = E[h(X'_T)]$. Since $h \in C_b(\partial A)$ is arbitrary, the laws of X_T, X'_T, must coincide. That is, for any CS there exists an SRS which yields the same exit distribution.

IV.3. The Finite Horizon Problem

We shall not give a separate treatment of the finite horizon control problem described by (C3) of section II.2. Instead we indicate how to reduce it to the problem of control up to an exit time and read off the relevant results from those already established above, after pointing out the few minor modifications needed.

This is achieved as follows. Let $S' = S \times \{0,1,2,...\}$ and L' be the countable product of replicas of D indexed by elements of S'. Define $A' \subset S'$ by $A' = S \times \{0,1,...,N-1\}$. Define the S'-valued process $\{X'_n\}$ by $X'_n = (X_n, n)$ for $n \geq 0$, $\{X_n\}$ being a controlled Markov chain as in the foregoing. Suppose that this $\{X_n\}$ is governed by a CS $\{\xi_n\}$. Define an L'-valued process $\{\xi'_n\}$ by: $\xi'_n((i, m)) = \xi_n(i)$ for $m = n$, = some fixed element of D otherwise. With these definitions, it is easy to see that $\{X'_n\}$ is an S'-valued controlled Markov chain governed by the CS $\{\xi'_n\}$ with its dynamics being given by

$$P(X'_{n+1} = (j, m)/X'_k, \xi'_k, k \leq n)$$

$$= \begin{cases} p(X_n, j, \xi_n(X_n)) & \text{when } m = n+1 \\ 0 & \text{otherwise.} \end{cases}$$

Define $k' : S' \times D \to R^+$, $h' : S' \to R^+$ by

$$k'((i, n), u)) = \iota(n, i, u),$$

$$h'((i, n)) = h(i)$$

for ι, h as in (II.2.4). Let $T = \min\{n \geq 0 \mid X'_n \notin A'\}$. The finite horizon cost (II.2.4) is now seen to equal

$$E\Big[\sum_{m=0}^{T-1} k'(X'_m, \xi'_m(X'_m)) + h'(X'_T)\Big] \tag{3.1}$$

which is of the type (II.2.2). Thus the problem of controlling $\{X_n\}$ on the finite horizon $\{0,1,..,N\}$ with cost (II.2.4) is equivalent to the problem of controlling $\{X'_n\}$ up to the first exit time T from A′ with cost (3.1).

Unlike A, A′ is not finite. However, the 'exit time' T is now uniformly bounded by N and thus has uniformly bounded moments (uniform with respect to the CS and the law of X'_0) of all orders. This also removes the need for assuming (∗). Except for minor technicalities to be described later, the developments of Sections 1 and 2 can now be mimicked step for step for the problem of controlling $\{X'_n\}$ to minimize (3.1). The obvious analogues of the foregoing results hold, which, after translation into the original terms of (C3), are stated below. Note in this context that an SRS (resp. SS) for $\{X'_n\}$ will correspond to an MRS (resp. MS) for $\{X_n\}$, since the second component of X'_n is simply the time count.

Consider the initial law $\pi \in \mathbb{P}(S)$ and let $A''_{CS}(\pi)$, $A''_{MRS}(\pi)$, $A''_{MS}(\pi)$ denote the sets of attainable $\nu_{F\pi}$ (recall (II.2.16)) over all CS, all MRS and all MS respectively.

Theorem 3.1. $A''_{CS}(\pi) = A''_{MRS}(\pi)$ and is compact convex. $A''_{MS}(\pi)$ is compact and contains the extreme points of $A''_{CS}(\pi)$. The latter correspond to MS $\{\xi_n\}$ for which a necessary condition is: $p(i,.,\xi_n(i))$ is an extreme point of $\{p_\Phi(i,.) \mid \Phi \in \mathbb{P}_0(L)\} \subset \mathbb{P}(S)$ for each $i \in S$, $n \geq 0$.

Theorem 3.2. An optimal MS $\{\bar{\xi}_n\}$ exists. This may be chosen so as to satisfy the following: For any n, $0 \leq n < N$, and any prescribed law π of X_n, $\{\bar{\xi}_m, n \leq m\}$ is optimal for the finite horizon control problem on the 'horizon' $\{n \leq m \leq N\}$ with the cost

$$E\Big[\sum_{m=n}^{N-1} \iota(m, X_m, \xi_m(X_m)) + h(X_N)\Big]. \tag{3.2}$$

We shall say that the above MS is optimal for any initial condition and any initial time. Define $V : S \times \{0,1,\ldots,N\} \to R$ by

$$V(i,N) = h(i)$$

and for $0 \leq n < N$, $V(i, n)$ is given by the expectation in (3.2) when $\xi_m = \bar{\xi}_m$, $m \geq n$, and X_m, $m \geq n$, is governed by $\{\bar{\xi}_m\}$ with $X_n = i$. Clearly,

$$V(i, n) = \min_{CS} E\Big[\sum_{m=n}^{N-1} \iota(m, X_m, \xi_m(X_m)) + h(X_N)/X_n = i\Big].$$

Theorem 3.3. V is the unique solution to the system of equations

$$V(i, n) = \min_u E\Big[\sum_{j \in S} p(i,j,u)V(j, n+1) + \iota(n, i, u))\Big], \quad 0 \leq n < N, \tag{3.3}$$

$$V(i, N) = h(i), \tag{3.4}$$

for $i \in S$, or alternatively,

$$V(i, n) = \min_\Phi \Big(\sum_{j \in S} \int p(i, j, u)\Phi(du)V(j, n+1) + \int \iota(n, i, u))\Phi(du)\Big) \tag{3.3'}$$

along with (3.4), for $i \in S$. Furthermore, an MS $\{\xi_n\}$ is optimal for any initial condition and any initial time if and only if for each i, n in (3.3), $\xi_n(i)$ attains the minimum on the right. (The obvious analogue for MRS also holds.) Also, $V : S' \to R^+$ is the unique solution to either of the above systems of equations.

There is one technicality here that needs to be taken care of. Recall the process $\{M_n\}$

defined in the proof of Theorem 2.2, where it was a zero mean bounded martingale. The corresponding process in the present case need not be so in view of the unboundedness of A' and the possible unboundedness of ι, h. Thus the application of the optional sampling theorem to this process as in Theorem 2.2 needs some modification. One replaces T by $T \wedge \tau_M$ where τ_M is the first exit time from $\{1,2,...,M\} \times \{0,1,...,N-1\} \subset S'$ for some $M \geq 1$ and applies the optional sampling theorem, letting $M \to \infty$ thereafter to arrive at the desired conclusions. The details are left to the reader.

The natural analogue of Theorem 2.4 for this set-up also holds. We omit the details.

A remark similar to that at the end of the preceeding section can be made here. Letting $\iota(.,.,.) \equiv 0$, one obtains the following result as an offshoot of the proof of Theorem 3.1. For any CS, there exists an MRS which leads to the same one dimensional marginals for the process. That is, given a chain $\{X_n\}$ governed by a CS $\{\xi_n\}$, there exists a chain $\{X'_n\}$ governed by an MRS such that the laws of X_n, X'_n coincide for each n. The proof of this statement is analogous to that of the corresponding statement for exit distributions at the end of the preceeding section.

Chapter V
Ergodic Control: Existence Results

With this chapter we begin our study of the 'ergodic' or 'long run average cost' control problem described by (C4) of section II.2. This chapter is devoted to establishing the existence of optimal SSS by direct methods and is based largely on [Bor 3], [Bor 4]. See also [Bor 2] for some earlier work and an application to queuing networks. We study two distinct set-ups, namely the so called near-monotone and stable cases. Assume for the time being that S is a single communicating class under all SRS.

V.1. Existence in the Near-Monotone Case

We start with some general results concerning invariant measures and limit behaviour of the joint empirical process for the state and control sequences. This leads in particular to the existence of an optimal SSS for near-monotone costs, a concept to be introduced later in this section.

Let $\{X_n\}$ be a chain governed by a CS $\{\xi_n\}$. Define the $\mathbb{P}(S \times D)$-valued empirical process $\{\nu_n, n \geq 1\}$ by

$$\nu_n(A \times B) = \frac{1}{n} \sum_{m=0}^{n-1} I\{X_m \in A, \ \xi_m(X_m) \in B\}, \ n \geq 1,$$

for A, B Borel in S, D respectively. Let $\bar{S} = S \cup \{\infty\}$ denote the one point compactification of S. By abuse of notation, we may identify ν_n with the element of $\mathbb{P}(\bar{S} \times D)$ that restricts to it on $S \times D$. Since $\mathbb{P}(\bar{S} \times D)$ is a compact space, $\{\nu_n\}$ viewed as a sequence of $\mathbb{P}(\bar{S} \times D)$-valued random variables converges to a sample path-dependent compact limit set in $\mathbb{P}(\bar{S} \times D)$. We characterize this set in Lemma 1.1 below, the statement of which calls for some new notation. Note that any element ν of $\mathbb{P}(\bar{S} \times D)$ can be decomposed as

$$\nu(A) = \delta_\nu \nu'(A \cap (S \times D)) + (1 - \delta_\nu)\nu''(A \cap (\{\infty\} \times D)) \quad (1.1)$$

for A Borel in $\bar{S} \times D$. $\delta_\nu \in [0, 1]$ is uniquely specified, $\nu' \in \mathbb{P}(S \times D)$ is uniquely specified if $\delta_\nu > 0$, and $\nu'' \in \mathbb{P}(\{\infty\} \times D)$ is uniquely specified if $\delta_\nu < 1$. We may render ν' and ν'' unique at all times by imposing an arbitrarily fixed choice thereof when $\delta_\nu = 0$ (resp. 1).

Lemma 1.1. Outside a set of zero probability, the following holds: For any limit point ν of

$\{\nu_n\}$ in $\mathbb{P}(\bar{S} \times D)$ for which $\delta_\nu > 0$,

$$\nu' = \nu_E[\Phi] \tag{1.2}$$

for some SSRS $\gamma[\Phi]$ with $\nu_E[\Phi]$ being the ergodic occupation measure defined by (II.2.19).

Proof. By the martingale stability theorem of [Lo], p. 53, we have

$$\lim_{n\to\infty} \frac{1}{n} \sum_{m=1}^{n} \left(I\{X_m = i\} - E[I\{X_m = i\}/\mathcal{F}_{m-1}] \right)$$

$$= \lim_{n\to\infty} \frac{1}{n} \sum_{m=1}^{n} \left(I\{X_m = i\} - \sum_{j\in S} p(j, i, \xi_{m-1}(j))I\{X_{m-1} = j\} \right)$$

$$= 0 \text{ a.s.}$$

for each $i \in S$. Consider a sample path outside the set of zero probability on which the above fails for any $i \in S$. Then for any ν as in the statement of the lemma, we must have

$$\nu'(\{i\} \times D) \geq \int p(.,i,.)d\nu' , \; i \in S.$$

After summing over $i \in S$ on both sides, both add up to one, and thus equality must hold in the above for each i. Disintegrate ν' as $\nu'(i, du) = \pi(i)\hat{\Phi}_i(du)$ for $\pi \in \mathbb{P}(S)$ and $i \to \hat{\Phi}_i : S \to \mathbb{P}(D)$. Set $\Phi = \Pi\hat{\Phi}_i$. Then the above translates into

$$\pi(i) = \sum_{j\in S} \pi(j)p_\Phi(j, i),$$

that is, $\pi = \pi[\Phi]$. The claim follows. QED.

Let $I_1 = \{\nu_E[\Phi] \mid \gamma[\Phi]$ an SSRS$\}$, $I_2 = \{\nu_E\{\xi\} \mid \gamma\{\xi\}$ an SSS$\}$ and $I_3 = \{\nu_E\{\xi\} \mid \gamma\{\xi\}$ an SSS and for each i, $p(i,.,\xi(i))$ is an extreme point of $\{p_\Phi(i,.) \mid \Phi \in \mathbb{P}_0(L)\} \subset \mathbb{P}(S)\}$.

Lemma 1.2. I_1, I_2 are closed, I_1 is convex and has its extreme points in $I_3 \subset I_2$.

Proof. Let $\Phi_n \in \mathbb{P}_0(L)$, $n \geq 1$, be such that $\gamma[\Phi_n]$ are SSRS and $\nu_E[\Phi_n] \to \nu$ for some $\nu \in \mathbb{P}(S \times D)$. Then for all $i \in S$,

$$\nu_E[\Phi_n](\{i\} \times D) = \int p(.,i,.)d\nu_E[\Phi_n], \; n \geq 1.$$

Letting $n \to \infty$, argue as above to conclude

$$\nu(\{i\} \times D) = \int p(.,i,.)d\nu.$$

Argue as in the proof of the preceding lemma to conclude that $\nu = \nu_E[\Phi]$ for some SSRS $\gamma[\Phi]$. This proves that I_1 is closed. The proof that I_2 is closed is similar. Convexity of I_1 follows as in Theorem III.1.2 with the system

$$\pi[\Phi]\, P[\Phi] = \pi[\Phi] \tag{1.3}$$

$$\pi[\Phi]1_c = 1 \tag{1.4}$$

(with $1_c \overset{\Delta}{=} [1,1,1,...]^T$) playing the role of (III.1.9). That the extreme points of I_1 lie in I_3 also follows as in Theorem III.1.3, once again with (1.3), (1.4) playing the role of (III.1.9).

QED.

In the above proof, let $\Phi' = \prod_i \hat{\Phi}'_i$ be a limit point of $\{\Phi_n\}$ in $\mathbb{P}_0(L)$. By dropping to a subsequence if necessary, let $\Phi_n \to \Phi'$. Then $P[\Phi_n] \to P[\Phi']$ termwise. Also, since $\nu_E[\Phi_n] \to \nu_E[\Phi]$, we have $\pi[\Phi_n] \to \pi[\Phi]$ in $\mathbb{P}(S)$ and hence in total variation. Thus $\pi[\Phi]P[\Phi'] \leftarrow \pi[\Phi_n]\, P[\Phi_n] = \pi[\Phi_n] \to \pi[\Phi]$, implying that $\pi[\Phi] = \pi[\Phi']$. Furthermore, for any $f \in C_b(S \times D)$,

$$\int f(i, u)\, \hat{\Phi}_{ni}(du) \to \int f(i, u)\, \hat{\Phi}'_i(du), \quad i \in S.$$

Since $\pi[\Phi_n] \to \pi[\Phi']$ in total variation,

$$\begin{aligned}\int f(i, u)d\nu_E[\Phi_n] &= \sum_{i \in S} \pi[\Phi_n](i) \int f(i, u)\, \hat{\Phi}_{ni}(du) \\ &\to \sum_{i \in S} \pi[\Phi'](i) \int f(i, u)\, \hat{\Phi}_i(du) \\ &= \int f d\nu_E[\Phi'].\end{aligned}$$

Hence we may take $\Phi = \Phi'$ in the above lemma for any limit point Φ' of $\{\Phi_n\}$. Note that these arguments require only that $\{\nu_E[\Phi_n]\}$ have a limit point in $\mathbb{P}(S \times D)$. Thus we have the following spin-off of the foregoing.

Corollary 1.1. If I_1 is tight, the map $\Phi \to \nu_E[\Phi]$ is continuous on the set $\{\Phi \in \mathbb{P}_0(L) \mid \gamma[\Phi]$ an SSRS$\}$.

An analogous result holds for I_2, the map $\xi \to \nu_E\{\xi\}$ and the set $\{\xi \in L \mid \gamma\{\xi\}$ an SSS$\}$ respectively. In fact the tightness of I_1 implies that of I_2, and vice versa, by virtue of the above lemma.

Corollary 1.2. If an optimal SSRS exists, an optimal SSS also exists and may be taken to be such that the corresponding $\nu_E\{\xi\}$ is in I_3.

The proof is analogous to that of Theorem III.1.4(a). Some care, however, is needed in the application of Choquet's theorem here since I_1 need not be compact. Consider I_1 as a subset of $\mathbb{P}(\bar{S} \times D)$ by identifying its elements with their unique extensions in the latter. Let $\bar{I}_1$ denote its closure, a compact convex set. By Choquet's theorem, each element μ of I_1 will be the barycenter of a probability measure m supported on the set of extreme points of $\bar{I}_1$. Now, each extreme point of I_1 must be an extreme point of $\bar{I}_1$ because it could not be otherwise without assigning a strictly positive probability to $\{\infty\} \times D$. If m assigns strictly positive probability to the extreme points of $\bar{I}_1$ which are not extreme points of I_1, then μ must assign a strictly positive probability to $\{\infty\} \times D$, which is false. Thus m must be supported on the set of extreme points of I_1 itself. This takes care of the problem caused by the possible lack of compactness of I_1.

We now state the near-monotonicity condition on the cost that ensures the existence of an optimal SSRS. Let

$$\beta = \inf_{\text{SSRS}} \int k d\nu_E[\Phi],$$

assumed to be finite if the problem is to make sense (recall (1.2)). Say that k is near-monotone if

$$\liminf_{i \to \infty} \min_u k(i, u) > \beta. \tag{1.5}$$

Intuitively, a near-monotone k discourages unstable behaviour by penalizing numerically large values of the state. (1.5) is satisfied by any k of the type $k(i, u) = f(i)$, $i \in S$, for a strictly increasing function $i \to f(i)$. This is the rationale for the term 'near-monotone'.

Theorem 1.1. If k is near-monotone, an optimal SSS exists.

Proof. In view of Corollary 1.2, it suffices to show that an optimal SSRS exists. Let $\gamma[\Phi_n]$, $n \geq 1$, be SSRS such that

$$\int k d\nu_E[\Phi_n] \downarrow \beta.$$

By identifying $\nu_E[\Phi_n]$ with the element of $\mathbb{P}(\bar{S} \times D)$ that restricts to it on $S \times D$ for each n, and then dropping to a subsequence if necessary, we may assume that $\nu_E[\Phi_n] \to \nu$ in

$\mathbb{P}(\bar{S} \times D)$ for some ν. Let $n \to \infty$ in the equation

$$\nu_E[\Phi_n](\{j\} \times D) = \int p(.,j,.)d\nu_E[\Phi_n], \; j \in S,$$

and argue as in Lemma 1.1 to conclude that for ν' as in (1.1), $\delta_\nu > 0$ implies

$$\nu'(\{j\} \times D) = \int p(.,j,.)d\nu', \; j \in S.$$

Disintegrating ν' as $\nu'(i, du) = \pi(i)\hat{\Phi}_i(du)$, $i \in S$, and setting $\Phi = \prod_i \hat{\Phi}_i$, we have $\pi = \pi[\Phi]$ and therefore $\nu' = \nu_E[\Phi]$. Let $k_m = k \wedge m$ for $m \geq 1$ and pick $\varepsilon > 0$ such that (1.5) continues to hold with $\beta + \varepsilon$ in place of β. Then

$$\beta = \lim_{n \to \infty} \int k d\nu_E[\Phi_n] \geq \lim_{n \to \infty} \int k_m d\nu_E[\Phi_n]$$

$$\geq \delta_\nu \int k_m \, d\nu_E[\Phi] + (1 - \delta_\nu)\,((\beta + \varepsilon) \wedge m).$$

Letting $m \to \infty$ on the right hand side,

$$\beta \geq \delta_\nu \int k d\nu_E[\Phi] + (1 - \delta_\nu)\,(\beta + \varepsilon)$$

$$\geq \delta_\nu \beta + (1 - \delta_\nu)\,(\beta + \varepsilon).$$

This is possible only if $\delta_\nu = 1$ and $\int k d\nu_E[\Phi] = \beta$. The only step that now remains is to show that under an arbitrary CS $\{\xi_n\}$,

$$\liminf_{n \to \infty} \frac{1}{n} \sum_{m=0}^{n-1} k(X_m, \xi_m(X_m)) \geq \beta \text{ a.s.} \tag{1.6}$$

For any sample point for which the conclusions of Lemma 1.1 hold, let ν be a limit point of $\{\nu_n\}$ in $\mathbb{P}(\bar{S} \times D)$ along some subsequence of $\{n\}$, denoted $\{n\}$ again by abuse of notation. For ν' defined as in (1.1), we have

$$\lim_{n \to \infty} \int k d\nu_n \geq \lim_{n \to \infty} \int k_m \, d\nu_n$$

$$\geq \delta_\nu \int k_m \, d\nu' + (1 - \delta_\nu)\,(\beta + \varepsilon) \wedge m).$$

Note that $\int k d\nu' \geq \beta$ in view of the definition of β and Lemma 1.1. Let $m \to \infty$ on the right hand side of the above inequality to deduce

$$\lim_{n\to\infty} \int k dv_n \geq \beta.$$

Since this is true for any limit point v of $\{v_n\}$ in $\mathbb{P}(\bar{S} \times D)$ and for all sample points outside a set of zero probability (in view of Lemma (1.1)), (1.6) follows. QED.

Recall that our original formulation of the ergodic control problem would require (1.6) to hold with limsup in place of liminf. Thus we have obtained a stronger claim than the one we sought.

V.2. Existence in the Stable Case

This section gives an existence result similar to the one above under a condition that ensures a kind of uniform stability under all SRS. This condition is not always easy to verify. However, we give in the next section some sufficient conditions for it which are more accessible. Consider the following.

Condition A. The stopping time $\tau = \min\{n > 0 \mid X_n = 1\}$ ($= \infty$ if $X_n \neq 1$ for all $n > 0$) with X_0 fixed at 1, is uniformly integrable under all SS.

In particular, this implies that all SS are SSS.

Lemma 2.1. The following are equivalent:

(i) Condition A holds.
(ii) $\{v_E\{\xi\} \mid \xi \in L\}$ is tight in $\mathbb{P}(S \times D)$.
(iii) $\{v_E\{\xi\} \mid \xi \in L\}$ is compact in $\mathbb{P}(S \times D)$.

Proof. Clearly (iii) implies (ii). Also, (ii) implies (iii) in view of the remarks following Corollary 1.1 and the compactness of L. Let (i) hold. Let $\xi^n \to \xi^\infty$ in L, and X^n_m, $m \geq 0$ be the corresponding chains governed by $\gamma\{\xi^n\}$, with $X^n_0 = 1$, $n = 1,2,\ldots,\infty$. By Corollary II.3.2, $[X^n_0, X^n_1,\ldots] \to [X^\infty_0, X^\infty_1,\ldots]$ in law as S^∞-valued random variables. By Skorohod's theorem, they may be assumed to be defined on a common probability space and the convergence to be a.s. Let $\tau^n = \min\{m > 0 \mid X^n_m = 1\}$, $n = 1,2,\ldots,\infty$. Then $\tau^n \to \tau^\infty$ a.s., and for any $i \in S$,

$$\sum_{m=0}^{\tau^n - 1} I\{X^n_m = i\} \to \sum_{m=0}^{\tau^\infty - 1} I\{X^\infty_m = i\} \text{ a.s.}$$

By Condition A, we can take expectations in the above to conclude that $E[\tau^n] \to E[\tau^\infty]$ and

$$E\Big[\sum_{m=0}^{\tau^n-1} I\{X_m^n = i\}\Big] \to E\Big[\sum_{m=0}^{\tau^\infty-1} I\{X_m^\infty = i\}\Big].$$

Thus

$$\pi\{\xi^n\}(i) = E\Big[\sum_{m=0}^{\tau^n-1} I\{X_m^n = i\}\Big] / E[\tau^n] \to$$

$$E\Big[\sum_{m=0}^{\tau^\infty-1} I\{X_m^\infty = i\}\Big] / E[\tau^\infty] = \pi\{\xi^\infty\}(i), \ i \in S.$$

Thus $\pi\{\xi^n\} \to \pi\{\xi^\infty\}$ in total variation by Scheffe's theorem and therefore in $\mathbb{P}(S)$. This proves that the map $\xi \in L \to \pi\{\xi\} \in \mathbb{P}(S)$ is continuous. Since L is compact, $\{\pi\{\xi\} \mid \xi \in L\}$ is compact and therefore tight in $\mathbb{P}(S)$. Since D is compact and $\pi\{\xi\}$ is the marginal in $\mathbb{P}(S)$ of $\nu_E\{\xi\} \in \mathbb{P}(S \times D)$, (ii) follows. Conversely, suppose (ii) holds and (i) fails. Recall that for non-negative integrable random variables, a.s. convergence implies convergence in the mean if and only if they are uniformly integrable, which in turn holds if and only if their norms converge. Thus if (i) fails, there must exist $\xi^n \to \xi^\infty$ in L such that for $\{X_m^n\}$, τ^n defined correspondingly as above, $\liminf E[\tau^n] > E[\tau^\infty]$. (The $\geq$ inequality would hold in any case by Fatou's lemma.) Since $\pi\{\xi^m\}(1) = (E[\tau^m])^{-1}$ for each m, we have $\limsup \pi\{\xi^m\}(1) < \pi\{\xi^\infty\}(1)$. Now for each $N \geq 1$,

$$\sum_{i=1}^{N} \pi\{\xi^m\}(i) p(i, j, \xi^m(i)) \leq \pi\{\xi^m\}(j).$$

Let π be a limit point of $\{\pi\{\xi^m\}\}$ in $\mathbb{P}(S)$. Passing to the limit along an appropriate subsequence in the above inequality, we have

$$\sum_{i=1}^{N} \pi(i) p(i, j, \xi^\infty(i)) \leq \pi(j).$$

Letting $N \to \infty$,

$$\sum_{i \in S} \pi(i) p(i, j, \xi^\infty(i)) \leq \pi(j).$$

Since both sides add up to one when summed over j, equality must hold for all j. Thus $\pi(1) = \pi\{\xi^\infty\}(1) = \lim \pi\{\xi^n\}(1)$, a contradiction. Therefore (i) must hold. This completes the proof. QED.

Corollary 2.1. Under condition A, all SRS and SSRS.

Proof. Let $\gamma[\Phi]$, $\Phi = \prod_i \hat{\Phi}_i$, be an SRS such that for some $i_0 \in S$ (say, $i_0 = 1$), $\hat{\Phi}_{i_0} = \hat{\Phi}_1 = a\varphi_1 + (1 - a)\varphi_2$ for some $\varphi_1, \varphi_2 \in \mathbb{P}(D)$ and $a \in (0,1)$, and the SRS $\gamma[\Phi']$, $\gamma[\Phi'']$ defined by $\Phi' = \varphi_1 \times \prod_{i=2}^{\infty} \hat{\Phi}_i$, $\Phi'' = \varphi_2 \times \prod_{i=2}^{\infty} \hat{\Phi}_i$ are SSRS. Letting $b \in (0,1)$ be such that

$$a = b\pi[\Phi'](1)/(b\pi[\Phi'](1) + (1 - b)\pi[\Phi''](1)),$$

argue as in the proof of Theorem III.1.3 (with (1.3) and (1.4) playing the role of (III.1.9)) to conclude that $\nu_E[\Phi] = b\nu_E[\Phi_1] + (1 - b)\nu_E[\Phi_2]$, proving *en passant* that $\gamma[\Phi]$ is an SSRS. Since all SS are SSS under condition A, this argument shows that all SRS $\gamma[\Phi]$, $\Phi = \prod_i \hat{\Phi}_i$, for which all but one $\hat{\Phi}_i$ are Dirac measures and the one that is not is a convex combination of two distinct Dirac measures, are SSRS. Iterating the argument, we claim the same when all but one $\hat{\Phi}_i$ are Dirac measures and the exceptional one has a finite support in D. Iterate the argument once more to conclude the same when all but finitely many $\hat{\Phi}_i$'s are Dirac measures and the rest have finite supports. Let $\mathbb{P}_1(L)$ denote the set of such Φ. Then $\mathbb{P}_1(L)$ is dense in $\mathbb{P}_0(L)$. For any given $\Phi \in \mathbb{P}_0(L)$ we can find a sequence $\{\Phi_n\}$ in $\mathbb{P}_1(L)$ converging to Φ in $\mathbb{P}_0(L)$. By the foregoing, $\{\gamma[\Phi_n]\}$ are SSRS and $\{\nu_E[\Phi_n]\}$ are in the convex hull of $\{\nu_E\{\xi\} \mid \xi \in L\}$ and therefore tight. Letting ν denote any limit point of $\nu_E[\Phi_n]$, $n \geq 1$, argue as in the proof of Corollary 1.1 to conclude that $\nu = \nu_E[\Phi]$. In particular, $\gamma[\Phi]$ is an SSRS. QED.

Corollary 2.2. The following are equivalent to each other and to (i) - (iii) of Lemma 2.1.

(iv) Condition A holds with 'SS' replaced by 'SRS'.

(v) $\{\nu_E[\Phi] \mid \Phi \in \mathbb{P}_0(L)\}$ is tight in $\mathbb{P}(S \times D)$.

(vi) $\{\nu_E[\Phi] \mid \Phi \in \mathbb{P}_0(L)\}$ is compact in $\mathbb{P}(S \times D)$.

Proof. The equivalence of (iv)-(vi) is proved along the lines of Lemma 2.1. Since all SS are SRS, (iv) implies (i). That (ii) implies (v) is contained in the proof of the preceding corollary. QED.

Corollary 2.3. There exists an $\gamma\{\xi\}$ for which $\pi\{\xi\} \in I_3$ and

$$\int k d\nu_E\{\xi\} = \beta.$$

Proof. This is immediate from (vi) above and Lemma 1.2 by arguments analogous to those leading to Theorem III.1.4(a). QED.

This guarantees the existence of an SSS that is optimal among all SRS. However, its optimality under all CS is not assured. We prove that under a stronger condition, namely condition B below, we do indeed have an SSS which is optimal among all CS.

Condition B. Letting $\tau = \min\{n > 0 \mid X_n = 1\}$ as before,

$$\sup_{CS} E[\tau^2 / X_0 = 1\} < \infty. \tag{2.1}$$

Consider a chain $\{X_n\}$ governed by an arbitrary CS $\{\xi_n\}$ with $X_0 = 1$. Define the stopping times

$$\tau_0 = 0,$$

$$\tau_{n+1} = \min\{m > \tau_n \mid X_m = 1\}, \ n \geq 0.$$

By (2.1) and Theorem II.3.1, it follows that under condition B,

$$\sup_n E[(\tau_{n+1} - \tau_n)^2] < \infty. \tag{2.2}$$

In particular, $\tau_n < \infty$ a.s. for all n.

Lemma 2.2. Under condition B, δ_ν in the statement of Lemma 1.1 may be taken to be one.

Remark. This is equivalent to saying that under condition B, outside a set of zero probability, the $\{\nu_n\}$ of the preceding section form a tight sequence.

Proof. By (2.2) and the martingale stability theorem of [Lo], p. 53, it follows that for any $N \geq 1$,

$$\lim_{n\to\infty} \frac{1}{n} \sum_{i=0}^{n-1} \Big(\big(\sum_{m=\tau_i}^{\tau_{i+1}-1} I\{X_m \geq N\} \big) - E\big[\sum_{m=\tau_i}^{\tau_{i+1}-1} I\{X_m \geq N\} / \mathcal{F}_{\tau_i} \big] \Big) = 0 \text{ a.s.} \tag{2.3}$$

Thus

$$\limsup_{n\to\infty} \nu_n(\{N, N+1, \ldots\} \times D) = \limsup_{n\to\infty} \frac{1}{n} \sum_{m=0}^{n-1} I\{X_m \geq N\}$$

$$\leq \limsup_{n\to\infty} \frac{1}{n} \sum_{i=0}^{n-1} \Big(\sum_{m=\tau_i}^{\tau_{i+1}-1} I\{X_m \geq N\} \Big)$$

$$= \limsup_{n\to\infty} \frac{1}{n} \sum_{i=0}^{n-1} E\Big[\sum_{m=\tau_i}^{\tau_{i+1}-1} I\{X_m \geq N\}/\mathcal{F}_{\tau_i}\Big]$$

$$\leq \sup_{CS} E\Big[\sum_{m=0}^{\tau-1} I\{X_m \geq N\}/X_0 = 1\Big], \tag{2.4}$$

in view of Theorem II.3.1. By an argument similar to the one leading to Theorem III.1.1, the supremum over all CS in (2.4) may be replaced by the supremum over all SRS. This in turn equals

$$\sup_{SRS\gamma[\Phi]} \Big(E[\tau/X_0 = 1] \big(\sum_{i \geq N} \pi[\Phi](i)\big) \Big).$$

In view of condition A (which is obviously implied by condition B) and (iv) - (v) of Corollary 2.2, the right hand side can be made smaller than any prescribed $\varepsilon > 0$ by choosing N sufficiently large. Let $\{\varepsilon_n\}$ be a sequence in (0,1) decreasing to zero. Consider a sample point for which the above holds for all $\varepsilon = \varepsilon_n$, $n \geq 1$. Such sample points have probability one, and for them, $\{\nu_n\}$ is a tight sequence. The claim follows. QED

Theorem 2.1. An optimal SSS exists.

Proof. This is immediate from Corollary 2.3, Lemma 1.1 and Lemma 2.2. In fact, one has, as in the last section, the stronger result

$$\liminf_{n\to\infty} \frac{1}{n} \sum_{m=0}^{n-1} k(X_m, \xi_m(X_m)) \geq \beta = \int k d\nu_E\{\xi\}$$

a.s. for $\gamma\{\xi\}$ as in Corollary 2.3. QED.

V.3. Sufficient Conditions for Stability

This section presents a few accessible conditions that imply condition A and/or condition B.

Suppose we need a bound of the type $\sup E[\tau^m/X_0 = 1] < \infty$ for some $m \geq 1$, the supremum being over all CS belonging to a particular class $\mathcal{A}$. One obvious way of ensuring this is to prove that for some $N \geq 1$, $K > 0$ and $0 < \varepsilon < 1$,

$$P(\tau \geq nN/X_0 = 1) \leq \varepsilon^n K \tag{3.1}$$

under all CS in $\mathcal{A}$. One can then argue as in the proof of Corollary IV.1.1 to obtain the desired bound. Many of the 'sufficient conditions' listed below are based on this idea.

(1) Suppose there exists an $N \geq 1$ such that

$$\sup_{CS} \sup_{i} P(\tau \geq N/X_1 = i) < 1.$$

Arguing as in Corollary IV.1.1, one obtains (3.1) with $\mathcal{A} = \{\text{all CS}\}$, implying condition B. If we replace the supremum over CS in the above by the supremum over SS, we similarly obtain condition A.

(2) Suppose that there exists a map $w : S \to R$ such that $w(i) \geq 1$ for all i, and furthermore,

(i) $\sup_{\xi} \sup_{1 \leq i \leq n} E[w(X_1)/X_0 = i] < \infty, \ n \geq 1,$

(ii) $\bigcup_{\xi \in L} \{i \in S \mid (w(i)/E[w(X_1)/X_0 = i]) \leq \ell\}$ is a finite set for all $\ell = 1,2,\ldots,$

the expectation in each case being under the SS $\gamma\{\xi\}$. As observed in [DoVa], p. 415, this ensures that all SS are SSS. Assume further that

$$c = \inf_{\xi} \pi\{\xi\}(i) > 0.$$

Let $C = \{\nu \in \mathbb{P}(S) \mid \nu(1) < c/2\}$ and $N > 4/c$. Then

$$\begin{aligned} P(\tau \geq N/X_0 = 1) &= P\Big(\frac{1}{N}\sum_{m=0}^{N-1} I\{X_m = 1\} \leq \frac{2}{N}/X_0 = 1\Big) \\ &\leq P\Big(\frac{1}{N}\sum_{m=0}^{N-1} I\{X_m = 1\} \leq c/2/X_0 = 1\Big) \\ &= P(\tilde{\nu}_N \in C/X_0 = 1) \end{aligned} \tag{3.2}$$

where $\tilde{\nu}_n \in \mathbb{P}(S)$ is the image of ν_n under the projection $S \times D \to S$. (That is,

$\tilde{\nu}_n(i) = n^{-1} \sum_{m=0}^{n-1} I\{X_m = i\}$ for $n \geq 1$, $i \in S$.).

Under the above hypotheses, the methods of [DoVa] yield an exponential bound of the type (3.1) for all SS. This can be verified by checking that the estimates of [DoVa], pp. 415–421, for the logarithm of the right hand side of (3.2) hold uniformly for all SSS under our hypotheses. Thus we have condition A. It seems plausible that this condition can be adapted so as to ensure condition B as well.

(3) (Uniform strong recurrence.) Suppose that

$$f(i) = \sup_{\xi} E[\tau / X_0 = i] < \infty, \quad i \in S, \tag{3.3}$$

and furthermore,

$$\sup_{\xi} E\left[\sum_{m=0}^{\tau-1} f(X_m) / X_0 = 1 \right] < \infty, \tag{3.4}$$

the expectation in both cases being under $\gamma\{\xi\}$. Under any CS,

$$E[\tau^2 / X_0 = 1] = 2E[\sum_{m=0}^{\tau-1} m / X_0 = 1] + E[\tau / X_0 = 1]. \tag{3.5}$$

Using (3.3) with $i = 1$ and adapting the argument of Theorem IV.1.1, one can prove

$$\sup_{CS} E[\tau / X_0 = 1] < \infty. \tag{3.6}$$

Also, using a similar argument and Theorem II.3.1 at the appropriate juncture,

$$\begin{aligned} E[\sum_{m=0}^{\tau-1} m / X_0 = 1] &= E[\sum_{m=1}^{\infty} (\tau - m) I\{\tau > m\} / X_0 = 1] \\ &= E[\sum_{m=1}^{\infty} E[(\tau - m) / \mathcal{F}_m] I\{\tau > m\} / X_0 = 1] \\ &\leq E[\sum_{m=0}^{\infty} f(X_m) I\{\tau > m\} / X_0 = 1] \\ &\leq E[\sum_{m=0}^{\tau-1} f(X_m) / X_0 = 1] \\ &\leq \sup_{CS} E[\sum_{m=0}^{\tau-1} f(X_m) / X_0 = 1]. \end{aligned} \tag{3.7}$$

By an argument analogous to that leading to Theorem IV.1.1, we may replace the supremum over CS in (3.7) by the supremum over SSS. Then by (3.4), (3.7) is finite. Together with (3.5) and (3.6) this implies condition B.

Note that if we assume

$$\sup_{SS} \sup_{i} E[\tau / X_0 = i] < \infty, \tag{3.8}$$

then (3.3) and (3.4) hold automatically. (3.8) is the classical uniform stability condition used for studying the ergodic control problem. See [FHT] for a discussion of this condition and its essential equivalence with several other conditions that have appeared in literature.

(4) (Liapunov condition) Suppose that there exist a $w : S \to R^+$, a finite subset A of S and an $\varepsilon > 0$ such that the following hold:

(i) $1 \in A$, $\partial A \overset{\Delta}{=} \{i \in S \backslash A \mid p(j, i, u) > 0$ for some $j \in A$ and $u \in D\}$ is finite.

(ii) $$\lim_{i \to \infty} w(i) = \infty, \tag{3.9}$$

(iii) Under any CS,

$$E[w(X_{n+1}) - w(X_n) + \varepsilon) I\{X_n \notin A\} / \mathcal{F}_n] \leq 0 \text{ a.s.} \tag{3.10}$$

(iv) There exists a random variable Z and a $\lambda > 0$ such that

$$E[\exp(\lambda Z)] < \infty, \tag{3.11}$$

and for all $c \in R$ and any CS,

$$P(| w(X_{n+1}) - w(X_n) | > c / \mathcal{F}_n) \leq P(Z > c). \tag{3.12}$$

Consider a chain $\{X_n\}$ governed by an arbitrary CS $\{\xi_n\}$ with $X_0 = i_0 \in A$. Define $\sigma_1 = \min\{n \geq 1 \mid X_n \notin A\}$ and $\sigma_2 = \min\{n > \sigma_1 \mid X_n \in A\}$.

Lemma 3.1. $$\sup_{i_0 \in A} \sup_{CS} E[\sigma_2^n] < \infty, \; n \geq 1.$$

Proof. Argue as in Corollary IV.1.1 to conclude that

$$\sup_{i_0 \in A} \sup_{CS} E[\sigma_1^n] < \infty, \; n \geq 1. \tag{3.13}$$

Let $\bar{\sigma} = \min\{n \geq 1 \mid X_n \in A\}$. Since ∂A is finite, it suffices (in view of (3.13) and Theorem II.3.1) to prove that

$$\sup_{CS} E[\bar{\sigma}^n / X_0 = i] < \infty, \quad i \in \partial A, \quad n \geq 1.$$

This follows from (2.15) of [Haj] with $Y_n = w(X_n)$. QED

Let $\{\tau_n\}$ denote the successive return times of $\{X_n\}$ to A with $\tau_0 = 0$. From the above lemma and Theorem II.3.1, we have $\tau_n < \infty$ a.s. for all n. Define an A-valued process $\{Z_n\}$ by

$$Z_n = X_{\tau_n}, \quad n \geq 0,$$

Let $\bar{\tau} = \min\{n \geq 1 | Z_n = 1\}$. Using the fact that S is a single communicating class under all SRS, one can adapt the arguments of Lemma IV.1.1 to conclude that there exist $N \geq 1$ and $\Delta \in (0,1)$ such that

$$\sup_{i \in A} \sup_{CS} P(\bar{\tau} > N / Z_0 = i) < 1 - \Delta.$$

Now mimick the arguments of Corollary IV.1.1 to conclude that for suitable $K > 0$, $a \in (0,1)$.

$$\sup_{i \in A} \sup_{CS} P(\bar{\tau} > n / Z_0 = i) < K^2 a^{2n}, \quad n \geq 1. \tag{3.14}$$

Lemma 3.2. Condition B holds.

Proof. Let $X_0 = Z_0 = 1$ in the foregoing. Then

$$\tau = \min\{n > 0 \mid X_n = 1\} = \sum_{m=0}^{\infty} (\tau_{m+1} - \tau_m) I\{\bar{\tau} > m\}. \tag{3.15}$$

By Lemma 3.1 and Theorem II.3.1.

$$\sup_i E[(\tau_{i+1} - \tau_i)^4] \leq (K')^2 < \infty \tag{3.16}$$

for a suitable constant $K' > 0$. Square both sides of (3.15) and take expectations. Using the Schwartz inequality along with (3.16),

$$E[\tau^2] \le \sum_{m=0}^{\infty} E[(\tau_{m+1} - \tau_m)^2 I\{\bar{\tau} > m\}]$$

$$+ 2 \sum_{n=0}^{\infty} \sum_{m<n} E[(\tau_{m+1} - \tau_m)(\tau_{n+1} - \tau_n) I\{\bar{\tau} > n\}]$$

$$\le K'K \sum_{m=0}^{\infty} a^m + 2K'K \sum_{m=0}^{\infty} ma^m < \infty$$

with a, K, K′ as in (3.14) and (3.16). QED

As an example, let $\infty > K > 2\varepsilon > 0$ and let $\{a_n, n \ge 1\}$ be a sequence in $[2\varepsilon, K]$. Consider a controlled Markov chain $\{X_n\}$ whose transition matrix P_u satisfies the following:

(i) $p(i, j, .) \equiv 0$ for $j \ne i+1$ or $i - 1$ and > 0 otherwise whenever $i \ne 1$, $p(1, j, .) \equiv 0$ for $j > 2$ and > 0 for $j \le 2$.

(ii) $p(i, i+1,.) \le (a_i - \varepsilon)/(a_i + a_{i+1})$, $i \in S$.

Let $w : S \to R$ be defined by: $w(1) = a_1$, $w(n)) = a_1 + \ldots + a_n$, $n \ge 2$. Then a straightforward computation shows that the Liapunov condition is satisfied for $A = \{1\}$.

V4. Miscellaneous Remarks

We conclude the chapter with some related remarks.

(1) Though condition B implies condition A, the converse is false, as the following example shows. Relabel S as $\{a_{00}, a_{10}, a_{11}, a_{20}, a_{21}, a_{22}, a_{30}, a_{31}, a_{32}, a_{33}, \ldots\}$. Let $D = \{1\}$ and $p(i, j, 1) = 1$ when $i = a_{mn}$. $j = a_{m(n+1)}$ with $m \ge 1, 0 \le n < m$,

$$= 1 \text{ when } i = a_{mm},\ m \ge 1,\ j = a_{00},$$

$$= Cm^{-3} \text{ when } i = a_{00}, j = a_{m0},\ m \ge 1.,$$

where $C^{-1} = \sum_{m=1}^{\infty} m^{-3}$. Let $\{X_n\}$ be a chain governed by these transition probabilities with $X_0 = a_{00}$ and let $\tau = \min\{n \ge 1 \mid X_n = a_{00}\}$. Then $2 \le \tau < \infty$ a.s. and $P(\tau = n) = C(n-1)^{-3}$ for $n \ge 2$. Hence $E[\tau] < \infty$ and $E[\tau^2] = \infty$.

However, it is tempting to conjecture the following:

(i) If all SS are SSS, condition A automatically holds.

(ii) Condition B can be dropped in the foregoing developments. (That is, condition A should suffice.)

(2) Suppose we drop the assumption that S is a single communicating class under all SRS. Suppose k is near-monotone. The arguments of the first section once again lead to the existence of an optimal SS $\gamma\{\xi\}$, such that $\int k d\nu_E\{\xi\} = \beta$. However, the support of $\pi\{\xi\}$ need not be the whole of S (though it may be taken to be a single communicating class). Thus everything is fine as long as the chain starts in B = support $(\pi\{\xi\})$. If not, suppose that for any initial law π the chain hits B in a finite time with probability one under some CS $\{\xi_n\}$, possibly depending on π. The optimal strategy then is to use $\{\xi_n\}$ until the chain hits B, and use $\gamma\{\xi\}$ thereafter. This situation occurs often in practice, e.g. in some controlled networks of queues when the state 'zero' (i.e., all queues empty) is attainable with probability one from any other state under any SS. When the above assumption also fails, the situation is much more complicated.

In the stable case, similar complications arise.

Chapter VI
Ergodic Control: Dynamic Programming

This chapter, which complements the preceding one, develops the dynamic programming equations for ergodic control leading to a characterization of optimal SSS. The material is largely from [Bor 3], [Bor 6] and [BoGh]. We continue to assume that S is a single communicating class under all SSRS.

VI.1 Preliminaries

We introduce the notation

$$C_f[\Phi] = \int f d\nu_E[\Phi],$$

$$C_f\{\xi\} = \int f d\nu_E\{\xi\}$$

for measurable $f: S \times D \to R$, whenever the right hand sides are well-defined. We shall develop the dynamic programming equations for ergodic control under the following blanket assumptions.

(i) S forms a single communicating class under all SRS.

(ii) An optimal SSS exists, and under any CS $\{\xi_m\}$,

$$\liminf_{n\to\infty} \frac{1}{n} \Sigma\, k(X_m, \xi_m(X_m)) \geq \beta \text{ a.s.}$$

(iii) (Stability under local perturbations.) If $\gamma\{\xi\}$ is an SSS satisfying $C_k\{\xi\} < \infty$, then for any $\xi' \in L$ such that $\xi'(i) \neq \xi(i)$ for at most one $i \in S$, $\gamma\{\xi'\}$ is an SSS and $C_k\{\xi'\} < \infty$.

Before proceeding further, let us see what this last condition entails. To start with, note that in the statement of (iii), 'at most one i' can be replaced by 'at most finitely many i'. If k is bounded, (iii) reduces to the following. If $\gamma\{\xi\}$ is an SSS and $\xi'(i) = \xi(i)$ for all but one i, then $\gamma\{\xi'\}$ is an SSS. Thus (iii) trivially holds in the stable case for bounded k. An important instance where (iii) holds is given by the following lemma.

Lemma 1.1. Suppose that for each $i \in S$, there exists a finite set $R_i \subset S$ such that $p(i,j,.) \equiv 0$ for $j \notin R_i$. Then (iii) holds.

Proof. Let $\gamma\{\xi\}$ be an SSS with $C_k\{\xi\} < \infty$ and $\{X_n\}$ a chain governed by $\gamma\{\xi\}$. Let $\tau(j) = \min\{m \geq 1 \mid X_n = j\}$, $j \in S$, with $\tau(j) = \infty$ if $X_n \neq j$ for all $n \geq 1$. Then

$$\infty > E[\tau(1)/X_0 = 1]$$

$$\geq P(\tau(j) < \tau(1)/X_0 = 1)E[\tau(1)/X_0 = j], \; j \in S,$$

by the strong Markov property. Since S is a single communicating class under $\gamma\{\xi\}$, $P(\tau(j) < \tau(1)/X_0 = 1) > 0$ for $j \neq 1$. Thus $a_j = E[\tau(1)/X_0 = j] < \infty$ for all $j \neq 1$ and hence for all j. Let $\xi' \in L$ be such that $\xi'(i) = \xi(i)$ for $i \neq 1$. Then under $\gamma\{\xi'\}$,

$$1 \leq a_1 = E[\tau(1)/X_0 = 1] \leq 1 + \sum_{j \in R_1 \setminus \{1\}} p(1, j, \xi'(1))a_j < \infty.$$

A similar argument using the fact that

$$E[\sum_{m=0}^{\tau(1)-1} k(X_m, \xi(X_m))/X_0 = 1] < \infty$$

under $\gamma\{\xi\}$ shows that under $\gamma\{\xi'\}$,

$$b = E[\sum_{m=0}^{\tau(1)-1} k(X_m. \xi'(X_m))/X_0 = 1] < \infty.$$

Thus $\gamma\{\xi'\}$ is an SSS with $C_k\{\xi'\} = b/a_1 < \infty$. QED

For an example where (iii) fails, consider S relabelled as in the last section of the preceding chapter. Let $D = [1.5,3]$ and the transition probabilities

$$p(i, j, u) = 1 \text{ for } u \in D, \; i = a_{mn}, j = a_{m(n+1)}, \; m \geq 1,$$

$$0 < n \leq m, \text{ or } i = a_{mm}, \; m \geq 1, \text{ and } j = a_{00},$$

$$= f(u)^{-1} m^{-u} \text{ for } u \in D, \; i = a_{00}, \; j = a_{m0}, \; m \geq 1,$$

where $f(u) = \sum_{n=1}^{\infty} n^{-u}$. Let $\{X_n\}$ be a Markov chain governed by an SS which picks the control u whenever it is in a_{00}. Letting $\tau = \inf\{n > 1 \mid X_n = a_{00}\}$, we have

$$E[\tau/X_0 = a_{00}] = f(u)^{-1} \sum_{m=1}^{\infty} (m+1) m^{-u}$$

which is finite for $u \in (2,3]$ and ∞ for $u \in [1.5,2]$.

Changing D to [2.5,3] in the above, one gets a situation where (iii) holds but there are infinitely many j for which $p(a_{00}, j, u) > 0$ for all u. (Take $j = a_{10}, a_{20}, \ldots$). Thus the condition of Lemma 1.1 is only a sufficient condition, not a necessary condition for (iii) to hold.

As an immediate consequence of our assumptions, we have the following.

Lemma 1.2. Let $\gamma\{\xi\}$ be an SSS with $C_k\{\xi\} < \infty$. Then for any $i \in S$, $u \in D$,

$$\sum_{j \in S} p(i, j, u)\, E_\xi [\sum_{n=0}^{\tau(1)} k(X_m, \xi(X_m))/X_0 = j] < \infty, \tag{1.1}$$

$$\sum_{j \in S} p(i, j, u)\, E_\xi [\tau(1)/X_0 = j] < \infty, \tag{1.2}$$

where $E_\xi[\,.\,]$ denotes the expectation under $\gamma\{\xi\}$.

Proof. Note that

$$\infty > E_\xi [\sum_{n=0}^{\tau(1)} k(X_n, \xi(X_n))/X_0 = 1]$$

$$\geq a E_\xi [\sum_{n=0}^{\tau(1)} k(X_n, \xi(X_n))/X_0 = j]$$

where $a = P(\tau(j) < \tau(1)/X_0 = 1)$ is strictly positive under $\gamma\{\xi\}$ as already observed. Thus

$$E_\xi [\sum_{n=0}^{\tau(1)} k(X_n . \xi(X_n))/X_0 = j] < \infty, \; j \in S\,. \tag{1.3}$$

Similarly,

$$E_\xi [\tau(1)/X_0 = j] < \infty, \; j \in S. \tag{1.4}$$

Let $\{\xi'_n\}$ denote a CS such that $\xi'_n = \xi$ for $n \geq 1$ and $\xi'_0(i) = u$ for prescribed $i \in S$, $u \in D$. Let $\{X'_n\}$ and $\{X_n\}$ be the chains governed by $\{\xi'_n\}$ and $\gamma\{\xi\}$ respectively, with $X'_0 = X_0 = i$. Let $\tau'(i) = \inf\{n \geq 1 \mid X'_n = i\}$. Then

$$k(i, u) + \sum_{j \in S} p(i, j, u)\, E_\xi [\sum_{n=0}^{\tau(1)} k(X_n, \xi(X_n))/X_0 = j]$$

$$= E[\sum_{n=0}^{\tau'(1)} k(X'_n, \xi'_n(X'_n))] + p(i, 1, u)(C_k\{\xi\}E_\xi[\tau(1)/X_0 = 1] - k(1, \xi(1))),$$

where we use the fact that

$$C_k\{\xi\} = E_\xi[\sum_{n=0}^{\tau(1)-1} k(X_n, \xi(X_n))/X_0 = 1]/E_\xi[\tau(1)/X_0 = 1].$$

The second term is clearly finite. The first term equals

$$E[\sum_{n=0}^{\tau'(1)} k(X'_n, \xi'_n(X'_n))\, I\{\tau'(1) < \tau'(i)\}]$$

$$+ \sum_{n=0}^{\tau'(1)} k(X'_n, \xi'_n(X'_n))\, I\{\tau'(1) > \tau'(i)\}].$$

Define $\rho \in L$ by $\rho(j) = \xi(j)$, $j \neq i$, and $\rho(i) = u$. Then the above is

$$\leq E_\rho[\sum_{n=0}^{\tau(1)} k(X_n, \rho(X_n))/X_0 = i]$$

$$+ E_\rho[\sum_{n=0}^{\tau(i)} k(X_n, \rho(X_n))/X_0 = i]$$

$$+ E_\xi[\sum_{n=0}^{\tau(1)} k(X_n, \xi(X_n))/X_0 = i] < \infty,$$

by virtue of (1.3) and the hypothesis of stability under local perturbation. (1.1) follows. (1.2) follows from (1.4) along similar lines. QED

VI.2 A Necessary Condition for Optimality

This section derives a necessary condition for the optimality of an SSS in terms of the dynamic programming equations (equations (2.2) below). Let $\gamma\{\xi\}$ be an SSS with $C_k\{\xi\} < \infty$. Define $V\{\xi\} = [V\{\xi\}(1), V\{\xi\}(2), ...]^T$ by

$$V\{\xi\}(i) = E_\xi[\sum_{n=0}^{\tau(1)-1} (k(X_n, \xi(X_n)) - C_k\{\xi\})/X_0 = i], \quad i \in S.$$

This is well defined by virtue of (1.3), (1.4) above. By Lemma 1.2,

$$\sum_{j \in S} p(i, j, u)V\{\xi\}(j), \ \ u \in D,$$

is also well-defined. Recall the definitions of $k[\Phi]$, $k\{\xi\}$ from Chapter III. Let I = the infinite identity matrix $[[\delta_{ij}]]$ and $1_c = [1,1,...]^T$.

Lemma 2.1. $V\{\xi\}(1) = 0$ and

$$C_k\{\xi\}1_c = (P\{\xi\} - I)\, V\{\xi\} + k\{\xi\}. \tag{2.1}$$

Proof. The first claim follows from the fact that

$$C_k\{\xi\} = \sum_{i \in S} \pi\{\xi\}(i)k(i, \xi(i))$$

$$= E_\xi\,[\sum_{n=0}^{\tau(1)-1} k(X_n, \xi(X_n))/X_0 = 1]/E[\tau(1)/X_0 = 1].$$

Since $V\{\xi\}(1) = 0$, one has

$$V\{\xi\}(i) = k(i, \xi(i)) - C_k\{\xi\} + E_\xi\,[\sum_{n=1}^{\tau(1)-1} (k(X_n, \xi(X_n)) -$$

$$C_k\{\xi\}))\, I\{\tau(1) > 1\}/X_0 = i]$$

$$= k(i, \xi(i)) - C_k\{\xi\} + E_\xi[V\{\xi\}(X_1)/X_0 = i]$$

$$= k(i, \xi(i)) - C_k\{\xi\} + \sum_{j \in S} p(i, j, \xi(i))\, V\{\xi\}(j)$$

for $i \in S$. (2.1) follows. QED

Let $A \subset S$ be a finite set containing a prescribed $i_0 \in S$ (say $i_0 = 1$) and $\xi' \in L$ be such that $\xi'(i) = \xi(i)$ for $i \in A$ with ξ as above. Let A_n, $n \geq 1$, be an increasing family of finite subsets of S containing A and increasing to S. Define

$$\sigma_m = \min\,\{n \geq 0 \mid X_n \notin A_m\}, \ \ m \geq 1,$$

$$\sigma = \min\{n \geq 0 \mid X_n \in A\}.$$

By our hypothesis of stability under local perturbation, $\gamma\{\xi'\}$ is an SSS with $C_k\{\xi'\} < \infty$.

Lemma 2.2. $\lim_{n\to\infty} E_{\xi'}[V\{\xi\}(X_{\sigma_n})I\{\tau(1) \geq \sigma_n\}/X_0 = 1] = 0.$

Proof. For $i \notin A$,

$$V\{\xi\}(i) = E_\xi [\sum_{n=0}^{\sigma-1} (k(X_n, \xi(X_n)) - C_k\{\xi\})/X_0 = i]$$

$$+ E_\xi [\sum_{n=\sigma}^{\tau(1)-1} (k(X_n, \xi(X_n)) - C_k\{\xi\})/X_0 = i].$$

The first term on the right is unchanged if $E_\xi[.]$ is replaced by $E_{\xi'}[.]$ and $k(X_n, \xi(X_n))$ by $k(X_n, \xi'(X_n))$. The second term is bounded in absolute value by

$$K = \max_{i \in A} E_\xi [\sum_{n=0}^{\tau(1)-1} (k(X_n, \xi(X_n)) + C_k\{\xi\})/X_0 = i].$$

Let $c = \max (C_k\{\xi\}, C_k\{\xi'\})$. Then for $i \notin A$,

$$| V\{\xi\}(i) | \leq E_{\xi'} [\sum_{n=0}^{\sigma-1} (k(X_n, \xi'(X_n)) + c)/X_0 = i] + K$$

$$\leq E_{\xi'} [\sum_{n=0}^{\tau(1)-1} (k(X_n, \xi'(X_n)) + c)/X_0 = i] + K.$$

Hence

$$| E_{\xi'}[V\{\xi\}(X_{\sigma_n})I\{\tau(1) \geq \sigma_n\}/X_0 = 1] |$$

$$\leq E_{\xi'} [\sum_{n=0}^{\tau(1)-1} (k(X_n, \xi'(X_n)) + c)I\{\tau(1) \geq \sigma_n\}/X_0 = 1]$$

$$+ K\, E_{\xi'}[I\{\tau(1) \geq \sigma_n]/X_0 = 1]$$

$\to 0$ as $n \to \infty$. QED

Theorem 2.1. If $\gamma\{\xi\}$ is an optimal SS,

$$\beta 1_c = \min_\varphi((P\{\varphi\} - I)V\{\xi\} + k\{\varphi\}), \tag{2.2}$$

or,

$$\beta 1_c = \min_{\Phi}((P[\Phi] - I)\, V\{\xi\} + k[\Phi]\,). \tag{2.2'}$$

Remark. By (2.1), the minimum in each case is attained respectively at $\varphi = \xi$, Φ = the Dirac measure at ξ.

Proof. We shall prove (2.2), the proof for (2.2′) being similar. Suppose (2.2) is false. Then there exist $i \in S$, $u \in D$ and $\Delta > 0$ such that for $\varphi \in L$ defined by $\varphi(j) = \xi(j)$, for $j \neq i$ and $\varphi(i) = u$, one has

$$\beta 1_c = (P\{\xi\} - I)V\{\xi\} + k\{\varphi\} + J \tag{2.3}$$

where $J = [0,0,\dots,0,\Delta,0,\dots,o]^T$ with the Δ being in the i'th place. Let $\{X_n\}$ be a chain governed by $\gamma\{\varphi\}$ with $X_0 = 1$. By our hypothesis of stability under local perturbation, $\gamma\{\varphi\}$ is an SSS with $C_k\{\varphi\} < \infty$. Set $A = \{i\}$ and $\{A_n\}$ as above. By (2.3),

$$\beta = E_\varphi[V\{\xi\}(X_{m+1})/X_m] - V\{\xi\}(X_m) + k(X_m, \varphi(X_m)) + \Delta I\{X_m = i\}$$

for $m \geq 0$. Thus for $n \geq 1$,

$$\beta(\tau(1) \wedge \sigma_n) = \sum_{m=0}^{\tau(1)\wedge\sigma_n - 1} (E_\varphi[V\{\xi\}(X_{m+1})/X_m] - V\{\xi\}(X_m))$$

$$+ \sum_{m=0}^{\tau(1)\wedge\sigma_n - 1} k(X_m, \xi(X_m)) + \Delta \sum_{m=0}^{\tau(1)\wedge\sigma_n - 1} I\{X_m = i\}. \tag{2.4}$$

Since $V\{\xi\}(X_{\tau(1)}) = V\{\xi\}(1) = 0$, we have

$$E_\varphi\Big[\sum_{m=0}^{\tau(1)\wedge\sigma_n - 1} (E_\varphi[V\{\xi\}(X_{m+1})/X_m] - V\{\xi\}(X_m))\Big]$$

$$= E_\varphi\Big[\sum_{m=0}^{\tau(1)\wedge\sigma_n} (V\{\xi\}(X_m) - E_\varphi[V\{\xi\}(X_m)/X_{m-1}]\,)\Big]$$

$$+ E[V\{\xi\}(X_{\sigma_n})\, I\{\tau(1) \geq \sigma_n\}\,]$$

$$= E_\varphi[V\{\xi\}(X_{\sigma_n})\, I\{\tau(1) \geq \sigma_n\}\,]$$

(by the optional sampling theorem),

$$\to 0$$

as $n \to \infty$ by Lemma 2.2. Taking expectations in (2.4), letting $n \to \infty$ and then dividing through by $E_\varphi[\tau(1)]$, we get

$$\beta = C_k\{\varphi\} + \Delta\pi\{\varphi\}(i) > C_k\{\varphi\},$$

contradicting the definition of β. The claim follows. QED

The function $i \in S \to V\{\xi\}(i) \in R$ for an optimal SSS $\gamma\{\xi\}$ is called the value function for the ergodic control problem. This definition depends upon our choice of the optimal SSS $\gamma\{\xi\}$ (which need not be unique) and upon which state we choose to label as '1'. In the remainder of this section, we essentially eliminate this dependence.

Lemma 2.3. Suppose $W = [W(1), W(2),...]^T$ satisfies

$$\beta 1_c = \inf_{\varphi} ((P\{\varphi\} - I)W + k\{\varphi\})$$

and for $V\{\xi\}$ as above

$$\sup_i |W(i) - V\{\xi\}(i)| < \infty$$

for some optimal SSS $\gamma\{\xi\}$. Then $W = V\{\xi\} + \text{constant} \times 1_c$. In particular, $W = V\{\xi\}$ if $W(1) = 0$.

Proof. Since

$$\beta 1_c = (P\{\xi\} - I)V\{\xi\} + k\{\xi\} \le (P\{\xi\} - I)W + k\{\xi\}$$

we have

$$(P\{\xi\} - I)(W - V\{\xi\}) \ge 0.$$

Thus under $\gamma\{\xi\}$, $W(X_n) - V\{\xi\}(X_n)$, $n \ge 0$, is a bounded submartingale with respect to the natural filtration of $\{X_n\}$ and hence converges a.s. Since $\{X_n\}$ visits each $i \in S$ infinitely often with probability one, this is possible only if $W(X_n) - V\{\xi\}(X_n)$, $n \ge 0$, is a.s. a constant sequence. The claim follows. QED

Lemma 2.4. Let $\gamma\{\xi\}$, $V\{\xi\}$ be as above. Define $V'\{\xi\} = [V'\{\xi\}(1), V'\{\xi\}(2),...]^T$ by

$$V'\{\xi\}(j) = E\,[\sum_{m=0}^{\tau(i)-1} (k(X_m, \xi(X_m)) - \beta)/X_0 = j],\ j \in S,$$

for a prescribed $i \in S$. Then

$$V'\{\xi\} = V\{\xi\} + \text{constant} \times 1_c.$$

Proof. For any $j \in S$,

$$|V\{\xi\}(j) - V'\{\xi\}(j)| \le E_\xi\,[\sum_{m=\tau(1)\wedge\tau(i)}^{\tau(1)\vee\tau(i)} (k(X_m, \xi(X_m)) + \beta)/X_0 = j\,]$$

$$\le E_\xi\,[\sum_{m=0}^{\tau(1)} (k(X_m, \xi(X_m)) + \beta)/X_0 = i\,] + E_\xi\,[\sum_{m=0}^{\tau(i)} (k(X_m, \xi(X_m)) + \beta)/X_0 = 1\,]. \quad (2.5)$$

Since the choice of state 1 in the definition of $V\{\xi\}$ was arbitrary, it is clear that (2.2) also holds with $V'\{\xi\}$ replacing $V\{\xi\}$. Also, the right hand side of (2.5) is finite by (1.3) and (1.4). The claim now follows from the preceding lemma. QED

Note that (2.2) is unchanged if we change $V\{\xi\}$ by a constant multiple 1_c. The next lemma eliminates the dependence of $V\{\xi\}$ on a specific choice of optimal $\gamma\{\xi\}$.

Lemma 2.5. For $i \in S$ and $\gamma\{\xi\}$, $V\{\xi\}$ as above,

$$V\{\xi\}(i) = \min E_\varphi[\sum_{m=0}^{\tau(1)-1} (k(X_m, \varphi(X_m)) - \beta)/X_0 = i]$$

where the minimum is over all SSS $\gamma\{\varphi\}$ (and is obviously attained at $\varphi = \xi$).

Proof. For $i = 1$, the claim follows from the fact that for any SSS $\gamma\{\varphi\}$,

$$E_\varphi\,[\sum_{m=0}^{\tau(1)-1} (k(X_m, \varphi(X_m)) - \beta)/X_0 = 1\,]$$

$$= E_\varphi[\tau(1)/X_0 = 1](C_k\{\varphi\} - \beta) \ge 0 = V\{\xi\}(1).$$

Take $i \neq 1$. Suppose the claim is false. Then for some SSS $\gamma\{\varphi\}$,

$$E_\varphi\,[\sum_{m=0}^{\tau(1)-1} (k(X_m, \varphi(X_m)) - \beta)/X_0 = i\,] < V\{\xi\}(i). \quad (2.6)$$

Consider the chain $\{X_n\}$ with $X_0 = 1$ and governed by a $CS\{\xi_n\}$ such that between each successive return to state 1, $\xi_n = \xi$ until $\{X_n\}$ hits i (if it does) and $= \varphi$ then on until it returns to 1. Under this CS,

$$E[\tau(1)/X_0 = 1] = E[\tau(1)I\{\tau(1) < \tau(i)\}/X_0 = 1]$$

$$+ E[\tau(i)I\{\tau(1) > \tau(i)\}/X_0 = 1]$$

$$+ E[(\tau(1) - \tau(i))I\{\tau(1) > \tau(i)\}/X_0 = 1]$$

$$\leq E_\xi[\tau(1)/X_0 = 1] + E_\xi[\tau(i)/X_0 = 1] + E_\varphi[\tau(1)/X_0 = i] < \infty.$$

Furthermore, the conditional law of $X_{\tau(1)}, X_{\tau(1)+1}, \ldots$ conditioned on $\mathcal{F}_{\tau(1)}$ is seen to be the same as that of $X_0, X_1, \ldots$. Let $\{\tau_i\}$ denote the successive return times to state 1 with $\tau_0 = 0$. Iterating the above argument, one has $E[\tau_i] < \infty$ for all i and for any measurable $f : S \times D \to R$, the sequence

$$\sum_{m=\tau_i}^{\tau_{i+1}-1} f(X_m, \xi_m(X_m)), \; i \geq 0,$$

is i.i.d. Now

$$E[\sum_{m=0}^{\tau(1)-1} (k(X_m, \varphi(X_m)) - \beta)]$$

$$= E_\xi[\sum_{m=0}^{\tau(1)-1} (k(X_m, \xi(X_m)) - \beta) I\{\tau(i) > \tau(1)\}]$$

$$+ E_\xi[\sum_{m=0}^{\tau(i)-1} (k(X_m, \xi(X_m)) - \beta) I\{\tau(i) < \tau(1)\}]$$

$$+ E[\sum_{m=\tau(i)}^{\tau(1)-1} (k(X_m, \xi(X_m)) - \beta) I\{\tau(i) < \tau(1)\}]$$

$$< \text{first two terms} + E_\xi[\sum_{m=\tau(i)}^{\tau(1)-1} (k(X_m, \xi(X_m)) - \beta)) I\{\tau(i) < \tau(1)\}]$$

$$= V\{\xi\}(1) = 0.$$

The strict inequality here follows by (2.6) and the fact that under $\gamma\{\xi\}$, $P(\tau(i) < \tau(1)/X_0 = 1) > 0$ (as already observed in the proof of Lemma 1.1). Since

$$\sum_{m=\tau_i}^{\tau_{i+1}-1} (k(X_m, \xi(X_m)) - \beta)), \ i \geq 1,$$

are i.i.d., the strong law of large numbers leads to

$$\frac{1}{\tau_n} \sum_{m=0}^{\tau_n - 1} (k(X_m, \xi(X_m)) - \beta)) \to E\Big[\sum_{m=0}^{\tau(1)-1} (k(X_m, \xi(X_m)) - \beta)/X_0 = 1 \Big] /$$

$$E[\tau(1)/X_0 = 1] < 0.$$

This contradicts the fact that under any CS $\{\xi_n\}$,

$$\liminf_{n \to \infty} \frac{1}{n} \sum_{m=0}^{n-1} k(X_m, \xi_m(X_m)) \geq \beta \text{ a.s.}$$

The claim follows. QED

VI.3. Sufficient Conditions for Optimality

In this section we develop sufficient conditions for the optimality of an SSS $\gamma\{\xi\}$ in terms of the dynamic programming equations. Let $\gamma\{\xi_0\}$ be an optimal SSS and $V\{\xi_0\}$ the value function.

Theorem 3.1. Suppose an SSS $\gamma\{\xi\}$ satisfies

$$\beta 1_c = (P\{\xi\} - I)V\{\xi_0\} + k\{\xi\}. \tag{3.1}$$

Then $\gamma\{\xi\}$ is optimal.

Proof. Let $\{\sigma_n\}$ be as in Lemma 2.2. Argue as in the proof of Theorem 2.1 to conclude that

$$\beta E_\xi[\tau(1) \wedge \sigma_n/X_0 = 1] = E_\xi\Big[\sum_{m=0}^{\tau(1)\wedge\sigma_n - 1} k(X_m, \xi(X_m))/X_0 = 1 \Big]$$

$$+ E_\xi [V\{\xi_0\}(X_{\sigma_n})I\{\tau(1) > \sigma_n)/X_0 = 1].$$

By Lemma 2.5, the last term is dominated by

$$E_\xi[E_\xi[\sum_{m=0}^{\tau(1)-\sigma_n-1} (k(X_{\sigma_n+m}, \xi(X_{\sigma_n+m})) - \beta))/X_{\sigma_n}]\, I\{\tau(1) > \sigma_n\}/X_0 = 1]$$

$$= E_\xi[\sum_{m=0}^{\tau(1)-\sigma_n-1} (k(X_{\sigma_n+m}, \xi(X_{\sigma_n+m})) - \beta)\, I\{\tau(1) > \sigma_n\}/X_0 = 1]$$

$$\to 0 \text{ as } n \to \infty.$$

Thus

$$\beta \geq E_\xi[\sum_{m=0}^{\tau(1)-1} k(X_m, \xi(X_m))/X_0 = 1]/E_\xi[\tau(1)/X_0 = 1]$$

$$= C_k\{\xi\}.$$

Since $\beta \leq C_k\{\xi\}$ in any case, $\beta = C_k\{\xi\}$ and the proof is complete. QED

For near-monotone k, we can replace SSS by SS in the statement of Theorem 3.1. In order to prove this, we shall need the following lemma.

Lemma 3.1. For near-monotone k, $\inf_i V\{\xi_0\}(i) > -\infty$.

Proof. By the near-monotonicity of k, the set $B = \{i \in S \mid k(i, \xi_0(i)) \leq \beta\} \cup \{1\}$ is finite. Let $\tau_B = \min\{n \geq 0 \mid X_n \in B\}$ ($= \infty$ if $X_n \notin B$ for all n). Then for $i \in S$ and $\{X_n\}$ governed by $\gamma\{\xi_0\}$ with $X_0 = i$, we have

$$V\{\xi_0\}(i) = E[\sum_{n=0}^{\tau(1)-1} (k(X_n, \xi_0(X_n)) - \beta)]$$

$$\geq E\Big[\sum_{n=\tau_B}^{\tau(1)-1} (k(X_n, \xi_0(X_n)) - \beta)\Big]$$

$$\geq \min_{j \in B} V\{\xi_0\}(j) \quad .$$ QED

Corollary 3.1. For near-monotone k, one can replace SSS by SS in the statement of Theorem 3.1.

Proof. Let $\gamma\{\xi\}$ be an SS ssatisfying (3.1). Let $A_n \subset S$, σ_n for $n \geq 1$ be as in Lemma 2.2. Arguing as in Theorem 2.1 for $\{X_n\}$ governed by $\gamma\{\xi\}$ with $X_0 = 1$, one has

$$\beta E[m \wedge \sigma_N] = E[V\{\xi_0\}(X_{m\wedge\sigma_N})]$$

$$- E\Big[\sum_{n=1}^{m\wedge\sigma_N} (V\{\xi_0\}(X_n) - E[V\{\xi_0\}(X_n)/\mathcal{F}_{n-1}])\Big]$$

$$+ E\Big[\sum_{n=0}^{m\wedge\sigma_N - 1} k(X_n, \xi(X_n))\Big].$$

By the optional sampling theorem, the second term on the right is zero. The preceding lemma then implies that the right hand side exceeds

$$- K + E\Big[\sum_{n=0}^{m\wedge\sigma_N - 1} k(X_n, \xi(X_n))\Big]$$

for some constant $K > 0$. Let $N \to \infty$, divide through by m and then let $m \to \infty$ to conclude that

$$\beta \geq \limsup_{m\to\infty} E\Big[\frac{1}{m}\sum_{n=0}^{m-1} k(X_n, \xi(X_n))\Big].$$

If $\gamma\{\xi\}$ is not an SSS, then $\{X_n\}$ is either null recurrent or transient. In either case, the measures $\{\nu_n\}$ defined as in section V.1 satisfy:

$$\lim_{n\to\infty} \nu_n(B \times D) = 0 \text{ for all finite } B \subset S.$$

In view of near-monotonicity of k, this holds in particular for the finite set B given by $B = \{i \in S \mid k(i, \xi_0(i)) \leq \beta + \varepsilon\}$ for some $\varepsilon > 0$ satisfying

$$\beta + \varepsilon < \liminf_{i\to\infty} \inf_u k(i, u).$$

Thus

$$\beta \geq \limsup_{m\to\infty} E\Big[\frac{1}{m}\sum_{n=0}^{m-1} k(X_n, \xi(X_n))\Big] \geq \beta + \varepsilon,$$

a contradiction. Thus $\gamma\{\xi\}$ must be an SSS. QED.

Next we obtain a variant of Theorem 3.1 along the lines of Theorem III.2.3. Call an SSS $\gamma\{\xi\}$ locally optimal if $C_k\{\xi\} < \infty$, and for all ξ' for which $\xi'(i)$, $\xi(i)$ differ for at most finitely many i, $C_k\{\xi\} \le C_k\{\xi'\}$.

Theorem 3.2. Suppose an SSS $\gamma\{\xi\}$ satisfies $C_k\{\xi\} < \infty$, and

$$C_k\{\xi\}1_c = \min_{\varphi}((P\{\varphi\} - I)V\{\xi\} + k\{\varphi\}). \tag{3.2}$$

Then it is locally optimal. Conversely, a locally optimal SSS $\gamma\{\xi\}$ must satisfy (3.2).

Proof. Let $\xi' \in L$ be such that $\xi'(i) \neq \xi(i)$ for at most finitely many i, say for $i \in A$ where $A \subset S$ is finite. By our hypothesis of stability under local perturbation, $\gamma\{\xi'\}$ is an SSS with $C_k\{\xi'\} < \infty$. Let $\{X_n\}$ be governed by $\gamma\{\xi'\}$ with $X_0 = 1$. By (3.2) and arguments analogous to those of Theorem 2.1, we obtain

$$C_k\{\xi\}E_{\xi'}[\tau(1) \wedge \sigma_n] \le E_{\xi'}\Big[\sum_{m=0}^{\tau(1)\wedge\sigma_n - 1} k(X_m, \xi'(X_m))\Big]$$

$$+ E_{\xi'}[V\{\xi\}(X_{\sigma_n})I\{\tau(1) \ge \sigma_n\}]$$

where the notation of Lemma 2.2 and Theorem 2.1 is used. Letting $n \to \infty$, it follows from Lemma 2.2 that

$$C_k\{\xi\}E_{\xi'}[\tau(1)] \le E_{\xi'}\Big[\sum_{m=0}^{\tau(1)-1} k(X_m, \xi'(X_m))\Big].$$

Divide through by $E_{\xi'}[\tau(1)]$ to obtain $C_k\{\xi\} \le C_k\{\xi'\}$. Conversely, let $\gamma\{\xi\}$ be locally optimal. Then (3.2) follows exactly as in Theorem 2.1. QED

Corollary 3.2. Suppose all locally optimal SSS are optimal. Then an SSS $\gamma\{\xi\}$ is optimal if and only if (3.2) holds.

Corollary 3.3. Suppose all SS are SSS and $\{\pi\{\xi\} \mid \xi \in L\}$ is tight in $\mathbb{P}(S)$. Suppose also that k is bounded. Then an SSS is optimal if and only if (3.2) holds.

Proof. Let $\gamma\{\xi_0\}$ be an optimal SSS and $\gamma\{\xi\}$ a locally optimal SSS. Define $\xi^n \in L$ by $\xi^n(i) = \xi_0(i)$ for $i \le n$ and $= \xi(i)$ for $i > n$, $n \ge 1$. Then $\xi^n \to \xi_0$ in L. From Corollary V.1.1, it follows that $\hat{\pi}\{\xi^n\} \to \hat{\pi}\{\xi_0\}$ and hence $C_k\{\xi^n\} \to C_k\{\xi_0\}$. But $C_k\{\xi^n\} \ge C_k\{\xi\}$ for each n by the local optimality of ξ. Thus $C_k\{\xi\} = C_k\{\xi_0\}$, and $\gamma\{\xi\}$ is optimal. The claim is now immediate in view of the preceding corollary. QED.

The problem of characterizing the exact conditions under which optimality and local optimality are equivalent is an interesting open problem.

VI.4 Characterization of the Value Function

By a solution of the dynamic programming equations (2.2), we mean a pair (c,W) with $c \in R^+$ and $W = [W(1),W(2),...]^T$ an infinite column vector, such that

$$c1_c = \min_{\xi} ((P\{\xi\} - I)W + k\{\xi\}). \tag{4.1}$$

Clearly the pair $(\beta,V\{\xi_0\})$ is one such solution when $\gamma\{\xi_0\}$ is an optimal SSS. Note that if (c,W) is one solution so is $(c,W + a1_c)$ for any $a \in R$. In this section we address the question of characterizing the distinguished solution $(\beta,V\{\xi_0\})$ from among all solutions. We do so in three cases: (i) the near-monotonicity condition on k, (ii) the Liapunov condition described in section V.3, and (iii) the uniform strong recurrence condition (V.3.3) with bounded k.

We start with (i). Thus k wili be assumed to be near-monotone. Let $G = \{f : S \to R \mid f(1) = 0, \inf_i f(i) > -\infty\}$.

Lemma 4.1. Let $(c,W) \in R^+ \times G$ satisfy (4.1). Then $c \geq \beta$.

Proof. Let $\varepsilon > 0$. By (4.1), there exists an SS $\gamma\{\xi\}$ such that

$$(c + \varepsilon)1_c \geq (P\{\xi\} - I)W + k\{\xi\}.$$

Let $\{X_n\}$ be governed by $\gamma\{\xi\}$ with $X_0 = 1$. We argue as in the proof of Corollary 3.1 to deduce that

$$c + \varepsilon \geq \limsup_{n \to \infty} E\Big[\sum_{m=0}^{n-1} k(X_m, \xi(X_m)) \Big] \geq \beta$$

where the second inequality must clearly hold whether or not $\gamma\{\xi\}$ is stable. Since $\varepsilon > 0$ is arbitrary, the claim follows. QED.

Lemma 4.2. If (β, W), $W \in G$, satisfies (4.1), then $W \geq V\{\xi_0\}$ termwise.

Proof. Let $\varepsilon > 0$ be such that

$$\liminf_{i \to \infty} \inf_u k(i, u) > \beta + \varepsilon . \tag{4.2}$$

By (4.1), there exists an SS $\gamma\{\xi\}$ such that

$$(\beta+\varepsilon)1_c \geq (P\{\xi\} - I)W + k\{\xi\}. \tag{4.3}$$

If $\gamma\{\xi\}$ is not an SSS, one may argue as in the proof of Corollary 3.1 to obtain a contradiction. Thus $\gamma\{\xi\}$ is an SSS. Let $\{X_n\}$ be governed by $\gamma\{\xi\}$ with $X_0 = i$ for some $i \in S$. For $\{\sigma_n\}$ as in Theorem 2.1, (4.3) leads to

$$\begin{aligned} W(i) \geq & \sum_{m=0}^{\tau(1)\wedge\sigma_n - 1} (k(X_m, \xi(X_m)) - (\beta+\varepsilon)) \\ & - \sum_{m=1}^{\tau(1)\wedge\sigma_n} (W(X_m) - E[W(X_m)/X_{m-1}]) \\ & + W(X_{\tau(1)\wedge\sigma_n}). \end{aligned} \tag{4.4}$$

Since $W \in G$, we have by Fatou's lemma,

$$\limsup_{n\to\infty} E[W(X_{\tau(1)\wedge\sigma_n})/X_0 = i] = \limsup_{n\to\infty} E[W(X_{\sigma_n})I\{\tau(1) > \sigma_n\}/X_0 = i] \geq 0$$

Taking expectations in (4.4) and letting $n \to \infty$, $\varepsilon \to 0$,

$$\begin{aligned} W(i) &\geq E_\xi\Big[\sum_{m=0}^{\tau(1)-1} (k(X_m, \xi(X_m)) - \beta)/X_0 = i\Big] \\ &\geq V\{\xi_0\}(i) \end{aligned}$$

by Lemma 2.5. QED

Corollary 4.1. For W as in the above lemma, $W = V\{\xi_0\}$.

Proof. Let $K > 0$ be such that $W(i) > -K$ for $i \in S$. Then for each i,

$$\sum_{j\in S} p(i, j, u)W(j) + k(i, u) = \sum_{j\in S} p(i, j, u)(W(j) + K) + k(i, u) - K.$$

The first term on the right is a monotone increasing limit of continuous functions in the variable u, and hence is lower semicontinuous in u. Since $k(i, .)$ is continuous, the left hand side is lower semicontinuous in u and hence attains a minimum at some $u_i \in D$. Let $\xi = [u_1, u_2, \ldots] \in L$. Then

$$\beta 1_c = (P\{\xi\} - I)W + k\{\xi\}.$$

By familiar arguments, it follows that $\gamma\{\xi\}$ is an SSS. Since for $\gamma\{\xi_0\}$ as before we have

$$\beta 1_c \le (P\{\xi\} - I)V\{\xi_0\} + k\{\xi\},$$

we have

$$(P\{\xi\} - I)(W - V\{\xi_0\}) \le 0.$$

Let $V = W - V\{\xi_0\}$. Letting $\{X_n\}$ be a chain governed by $\gamma\{\xi\}$ with $X_0 = 1$, it follows that for $\{\sigma_n\}$ as bove,

$$E[V(X_{n \wedge \sigma_m + 1})/X_{n \wedge \sigma_m}] \le V(X_{n \wedge \sigma_m}), \ m, n \ge 0.$$

By the above lemma.

$$V(X_n) \ge 0 = V(X_0).$$

The two inequalities together lead to $V(X_n) = 0$ a.s. for all $n \ge 0$. Since $\gamma\{\xi\}$ is an SSS, $X_n = i$ infinitely often a.s. for each $i \in S$. Thus $V(i) = 0$ for $i \in S$, implying the result.

QED.

The foregoing is summarized in the following theorem.

Theorem 4.1. Among all $(c,W) \in R^+ \times G$ satisfying (4.1), $(\beta, V\{\xi_0\})$ is the unique pair corresponding to the least value of c.

The following example shows that one cannot hope to improve on this in general. Let S be as above and consider the uncontrolled chain $\{X_n\}$ with transition probailities

$$p(1,1) = p(i, i-1) = 3/4, \ i \ge 2,$$

$$p(i, i+1) = \frac{1}{4}, i \ge 1.$$

Let $k \equiv 0$. Thus $\beta = 0$. Recall the example at the end of section V.3. The above is a special case thereof with $\varepsilon = 1$ and $a_n = 2$ for all n. Thus the chain is stable. (4.1) now reduces to

$$\frac{1}{4}W(2) + \frac{3}{4}W(1) - W(1) = c,$$

$$\frac{1}{4}W(i+1) + \frac{3}{4}W(i-1) - W(i) = c, \ i \ge 2.$$

Suppose that $W(1) = 0$. Then $W(2) = 4c$. The second equation above can be rewritten as

$$W(i+1) - W(i) = 4c + 3(W(i) - W(i-1)), \ i \geq 2.$$

Clearly for $c \geq 0$, $W(2) - W(1) \geq 0$. By using the above equation recursively, $W(i+1) - W(i) \geq 0$ for all i. Thus $W(i)$ increases with i and is in particular bounded from below. Thus for each $c \in R^+$, we can find a $W \in G$ such that (c,W) solves (4.1). The distinguished solution which we want is of course $(0,[0,0,...]) \in R^+ \times G$.

We next consider the Liapunov condition introduced in section V.3. Recall from there the 'Liapunov function' $w : S \to R^+$. Say that a function $f : S \to R$ is $O(w(.))$ if

$$\limsup_{i \to \infty} | f(i) | / w(i) < \infty.$$

Assume that k is bounded. Let the finite set $A \subset S$ and the scalar $\varepsilon > 0$ be as in the statement of the Liapunov condition. Let $\tau_A = \min\{n \geq 0 \mid X_n \in A\}$. Recall that under the Liapunov condition all SRS (resp. SS) are SSRS (resp. SSS).

Lemma 4.3. Under any SRS $\gamma[\Phi]$,

$$E[\tau_A / S_0 = i] \leq w(i)/\varepsilon, \ i \notin A.$$

Proof. Without any loss of generality, let $A = \{1,2,..., M\}$. Let $N > M$ and $M < i < N$. Define

$$\tau_N = \min\{n \geq 0 \mid X_n \notin \{M+1,...,N\}\}.$$

Since

$$\sum_{m=1}^{n \wedge \tau_N} (w(X_m) - E[w(X_m)/\mathcal{F}_{m-1}])$$

is a zero-mean martingale,

$$E[w(X_{n \wedge \tau_N + 1})/\mathcal{F}_{n \wedge \tau_N}] - w(i)$$

$$= - E\Big[\sum_{m=1}^{n \wedge \tau_N} (w(X_m) - E[w(X_m)/\mathcal{F}_{m-1}]) \Big] + E[w(X_{n \wedge \tau_N + 1})/\mathcal{F}_{n \wedge \tau_N}] - w(i)$$

$$= - E\Big[\sum_{m=0}^{n \wedge \tau_N} (E[w(X_{m+1})/X_m] - w(X_m))\Big]$$

$$\leq - \varepsilon\, E[n \wedge \tau_N].$$

Thus

$$E[n \wedge \tau_N] \leq w(i)/\varepsilon.$$

Let $n \to \infty$, then $N \to \infty$ and use Fatou's lemma to conclude. QED

Lemma 4.4. $V\{\xi_0\}$ is $O(w(.))$.

Proof. Let $i \notin A$. Then for $\{X_n\}$ governed by $\gamma\{\xi_0\}$,

$$V\{\xi_0\}(i) = E\Big[\sum_{m=0}^{\tau_A - 1} (k(X_m, \xi_0(X_m)) - \beta) + V\{\xi_0\}(X_{\tau_A})/X_0 = i\Big]$$

$$\leq 2K\, E[\tau_A/X_0 = i] + \sup_{j \in A} V\{\xi_0\}(j)$$

where K is a bound on k. The claim now follows from Lemma 4.3. QED

Lemma 4.5. Under any SRS $\gamma[\Phi]$, $\sum_i \pi[\Phi](i)w(i) < \infty$.

Proof. Consider (2.8) of [Haj] with $Y_k = w(X_k)$, $\{X_n\}$ being a chain governed by the SRS $\gamma[\Phi]$. Letting $k \to \infty$ in the resulting expression, one obtains

$$f(b) \overset{\Delta}{=} \sum_{\{i | w(i) \geq b\}} \pi[\Phi](i) \leq \frac{D}{1 - \rho}[e^{\eta(a-b)}], \quad b \geq 0,$$

where ρ, a, D, η are as in [Haj]. Thus

$$\sum_i \pi[\Phi](i)w(i) = \int_0^\infty f(b)db < \infty.$$ QED

Theorem 4.2. Under the Liapunov condition, $(\beta, V\{\xi_0\})$ is the unique solution to (4.1) in the class $\{(c,W) \mid c \in R,\ W \text{ is } O(w(.)),\ W(1) = 0\}$.

Proof. In view of the foregoing, $(\beta, V\{\xi_0\})$ is clearly a solution to (4.1) in the desired class. Let (c, W) be another solution. Then under any SS $\gamma\{\xi\}$,

$$c\,1_c \leq P\{\xi\}W - W + k\{\xi\}.$$

Moreover for any $\varepsilon > 0$, there exists an SS $\gamma\{\xi_\varepsilon\}$ such that

$$(c+\varepsilon)1_c \geq P\{\xi_\varepsilon\}W - W + k\{\xi_\varepsilon\}.$$

In view of Lemma 4.5, we shall left-multiply the first inequality by the row vector $\pi\{\xi\}$ and the second by the row vector $\pi\{\xi_\varepsilon\}$ to obtain (via Fubini's theorem)

$$\begin{aligned} c &\leq \sum_{i,j} \pi\{\xi\}(i)p(i, j, \xi(i))W(j) - \sum_i \pi\{\xi\}(i)W(i) + C_k\{\xi\} \\ &= C_k\{\xi\}; \end{aligned}$$

and analogously,

$$c + \varepsilon \geq C_k\{\xi_\varepsilon\}.$$

Since $\varepsilon > 0$ and $\xi \in L$ were arbitrary, it follows that $c = \beta$. Thus under an optimal SSS $\gamma\{\xi_0\}$,

$$\beta 1_c \leq (P(\xi_0\} - I)W + k\{\xi_0\},$$

$$\beta 1_c = (P\{\xi_0\} - I)V\{\xi_0\} + k\{\xi_0\}.$$

Together, these imply that $P\{\xi_0\}(W - V\{\xi_0\}) \geq W - V\{\xi_0\}$ termwise. Let $\{X_n\}$ be governed by $\gamma\{\xi_0\}$ with initial law $\pi\{\xi_0\}$. Then $\{X_n\}$ is a stationary process. By virtue of the foregoing, $W(X_n) - V\{\xi_0\}(X_n)$, $n \geq 1$, is a submartingale with respect to the natural filtration of $\{X_n\}$. Also,

$$\sup_n E[\,|W(X_n) - V\{\xi_0\}(X_n)|\,] = \sum_i \pi\{\xi_0\}(i)\,|W(i) - V\{\xi_0\}(i)|$$

$$< \infty$$

by Lemma 4.5 and the fact that W, $V\{\xi_0\}$ are $O(w(.))$. By the submartingale convergence theorem, $W(X_n) - V\{\xi_0\}(X_n)$ converges a.s. as $n \to \infty$. Since $\{X_n\}$ is positive recurrent, it visits each $i \in S$ infinitely often a.s. Thus the above convergence is possible if and only if $W(i) - V\{\xi_0\}(i)$ is a constant independent of i. Since $W(1) = 0 = V\{\xi_0\}(1)$, this constant must be zero. QED.

Recall the example above to show that Theorem 4.1 cannot in general be improved upon. The same example leads to identical remarks for Theorem 4.2 as well.

Finally, consider the uniform strong recurrence condition (V.3.8) with bounded k. From this and the definition of $V\{\xi_0\}$, it follows that $V\{\xi_0\}(i)$ is bounded uniformly in $i \in S$.

Theorem 4.3. Under the uniform strong recurrence condition (V.3.8) with bounded k, $(\beta, V\{\xi_0\})$ is the unique solution to (4.1) in the class $\{(c, W) \mid c \in R, W(1) = 0, \sup_i |W(i)| < \infty\}$.

Proof. This can be proved in exactly the same way as Theorem 4.2. QED

Chapter VII
Multiobjective Control Problems

The first section of this chapter discusses a multiobjective control problem recast as a constrained control problem. It is based on [Bor 7] which in turn builds upon the earlier work of [Be Ro], [Ross]. For sake of being specific, we take up the ergodic control set-up. Analogous developments for other criteria are possible: these will be only cursorily mentioned. The second section discusses Pareto-optimal and 'utopian' solutions of multi-objective problems, and sketches some other possible applications of the theory developed so far. This is based on [Gh]. See [AlSh] and the references therein for related works.

VII.1 Ergodic Control with Constraints

Recall the framework of the ergodic control problem. Let $k_i : S \times D \to R^+$, $1 \le i \le m$, be bounded continuous maps and β_i, α_i, $1 \le i \le m$, prescribed numbers with $\beta_i \le \alpha_i$ for each i. The control problem with constraints mentioned above is the problem of minimizing $\int k d\nu_E[\Phi]$ over all SSRS $\gamma[\Phi]$ under the constraints

$$\beta_i \le \int k_i \, d\nu_E[\Phi] \le \alpha_i, \quad 1 \le i \le m. \tag{1.1}$$

Let $G = \{\nu_E[\Phi] \mid \gamma[\Phi] \text{ an SSRS}\}$ and $G' \subset G$ the subset where (1.1) holds. We assume that G' is nonempty, the problem being vacuous otherwise. G' is clearly convex. We also assume that each $\{\nu \in G \mid \int k_i \, d\nu \ge \beta_i\}$ is closed. This is clearly the case when $\{k_i\}$ are bounded or when $\beta_i \le \min_{\nu \in G} \int k_i \, d\nu$. Finally, we assume that G is compact. Sufficient conditions for this to hold have already been discussed in Chapter V. In particular, this implies that G' will also be compact.

Our aim will be to show the existence of an optimal SSRS $\gamma[\Phi]$, $\Phi = \prod_i \hat{\Phi}_i$, such that $\hat{\Phi}_i$ is a Dirac measure for all i except for at most m, and is a convex combination of Dirac measures for these exceptional values of i. We also derive some necessary conditions for the optimal strategy. We shall proceed via several lemmas.

Let $f \in C(S \times D)$ be bounded from below such that the set $\{\nu \in G \mid \int f d\nu < \infty\}$ is non-empty. For some $a \in R$, let $H = G \cap \{\nu \mid \int f d\nu \le a\}$, assumed to be nonempty. Clearly, H is closed convex. Let $\nu_E[\Phi]$, $\Phi = \prod_i \hat{\Phi}_i$, be an extreme point of H. Suppose it is not an extreme point of G itself. Then there exist distinct measures $\nu_E[\Phi_{11}]$, $\nu_E[\Phi_{12}]$ such that at least one of them (say $\nu_E[\Phi_{11}]$) is not in H, and $\nu_E[\Phi]$ is a convex combination of the two. Suppose $\nu_E[\Phi_{21}] \in G\backslash H$, $\nu_E[\Phi_{22}]$ is another pair for which this holds.

In the next lemma and in what follows, it will help to view G and G′ as subsets of the Banach space of signed measures on $S \times D$, denoted $B(S \times D)$.

Lemma 1.1. $\nu_E[\Phi_{ij}]$, $1 \le i, j \le 2$, are colinear.

Proof. Suppose not. Clearly,

$$\infty > \int f d\nu_E[\Phi_{11}], \int f d\nu_E[\Phi_{21}] > \int f d\nu_E[\Phi] > \int f d\nu_E[\Phi_{12}], \int f d\nu_E[\Phi_{22}]. \tag{1.2}$$

Consider the rectangle formed by $\nu_E[\Phi_{ij}]$, $1 \le i, j < 2$, in $B(S \times D)$. The diagonals of the rectangle intersect at $\nu_E[\Phi]$. By (1.2), it is transversal to the hyperplane $\{\nu \in B(S \times D) \mid \int f d\nu = \int f d\nu_E[\Phi]\}$. Let $\nu_E[\Phi']$, $\nu_E[\Phi'']$ denote the points of intersection of the line joining $\nu_E[\Phi_{11}]$ with $\nu_E[\Phi_{22}]$ (resp., $\nu_E[\Phi_{21}]$ with $\nu_E[\Phi_{12}]$) with this hyperplane. (They must intersect by (1.2). Also, the intersections are in G by the convexity of G.) Then

$$\int f d\nu_E[\Phi'] = \int f d\nu_E[\Phi''] = \int f d\nu_E[\Phi].$$

Thus $\nu_E[\Phi']$, $\nu_E[\Phi'']$ are in H. Since the intersection of the rectangle and the hyperplane will be a line segment, $\nu_E[\Phi]$ is a convex combination of $\nu_E[\Phi']$ and $\nu_E[\Phi'']$.

This contradicts the fact that $\nu_E[\Phi]$ is an extreme point of H, proving the claim. QED.

Thus all pairs of points in G satisfying: (a) at least one of them is not in H, and (b) $\nu_E[\Phi]$ is a convex combination thereof, lie on a single straight line in $B(S \times D)$. Let Z denote the intersection of this line with G. Under our hypotheses on G, Z is a closed finite line segment. Let $\nu_E[\Phi_1]$, $\nu_E[\Phi_2]$ denote its end points.

Lemma 1.2. $\nu_E[\Phi_1]$, $i = 1,2$, are extreme points of G.

Proof. Suppose (say) $\nu_E[\Phi_i]$ is not. Then $\nu_E[\Phi_1]$ can be written as a convex combination of two distinct measures $\nu_E[\Phi_{11}], \nu_E[\Phi_{12}]$ in G. These cannot lie in Z because $\nu_E[\Phi_1]$ is an end point of Z. The claim follows by arguments similar to those of the preceding lemma, with the triangle formed by $\nu_E[\Phi_{11}]$, $\nu_E[\Phi_{12}]$, $\nu_E[\Phi_2]$ replacing the rectangle there. QED.

Now suppose that $\nu_E[\Phi]$ is not an extreme point of G. Then it lies in the relative interior of Z.

Corollary 1.1. For any $\nu \in Z$, $\int f d\nu$ is finite and ν is an extreme point of $H' = G \cap \{\mu \mid \int f d\mu \le \int f d\nu\}$.

Proof. The claim is trivial for $\nu = \nu_E[\Phi]$. Suppose $\nu \ne \nu_E[\Phi]$. Then there exists a $\nu' \in Z$ such that

$$\nu_E[\Phi] = b\nu' + (1 - b)\nu$$

for some $b \in (0,1)$. Thus

$$\int f d\nu_E[\Phi] = b\int f d\nu' + (1 - b)\int f d\nu < \infty.$$

The first claim follows. Suppose there exist distinct $\nu_1, \nu_2 \in H'$ such that $\nu = a'\nu_1 + (1-a')\nu_2$ for some $a' \in (0,1)$. Then by arguments analogous to those of the above lemmas, $\nu_E[\Phi]$ is seen to lie in the relative interior of the line segment formed by intersecting the hyperplane $\{\mu \mid \int f d\mu = \int f d\nu_E[\Phi]\}$ with the triangle formed by ν_1, ν_2 and ν', leading to a contradiction. The claim follows. QED

Let $\nu_E\{\xi_i\}$, $\xi_i = [\xi_i(1), \xi_i(2),...] \in L$, $i = 1,2$, be the end points of Z. We shall only consider the case when they are distinct. (If not, $\nu_E[\Phi]$ is an extreme point of G itself.) Thus for some $a' \in (0,1)$, $\nu_E[\Phi] = a'\nu_E\{\xi_1\} + (1-a')\nu_E\{\xi_2\}$. Arguing as in the proof of Theorem III.1.2, it is clear that we may take $\hat{\Phi}_i$ to be a convex combination of $\xi_1(i)$ and $\xi_2(i)$ for each i. (Compare with Lemma V.1.2.)

Let $\xi = [\xi(1), \xi(2),...] \in L$ be such that for each i, $\xi(i) =$ either $\xi_1(i)$ or $\xi_2(i)$.

Lemma 1.3. $\nu_E\{\xi\} \in Z$.

Proof. Suppose that $p(1,., \xi_1(1)) \neq p(1,., \xi_2(1))$. (If not, consider the smallest j for which $p(j,., \xi_1(j)) \neq p(j,., \xi_2(j))$.) Let φ_1, φ_2 denote the Dirac measures at $\xi_1(1), \xi_2(1)$ respectively. From the proof of Lemma V.1.2, it is clear that for $\Phi_1 = \varphi_1 \times \prod_{i=2}^{\infty} \hat{\Phi}_i$ and $\Phi_2 = \varphi_2 \times \prod_{i=2}^{\infty} \hat{\Phi}_i$, $\nu_E[\Phi_i]$, $i = 1,2$, are distinct and $\nu_E[\Phi]$ is different from either, and a convex combination of the two. By Lemma 1.1., $\nu_E[\Phi_i]$, $i = 1,2$, must lie on Z. Repeat the argument with Φ_1 or Φ_2 replacing Φ and state 2 in place of state 1 above. Using Corollary 1.1, it follows as above that for i,j = 1 or 2 and $\Phi' = \varphi_i \times \tilde{\varphi}_j \times \prod_{k=3}^{\infty} \hat{\Phi}_k$, $\nu_E[\Phi']$ is in Z, $\tilde{\varphi}_j$ being the Dirac measure at $\xi_j(2)$. Iterating the argument, it follows that for any $\Phi' = \prod_i \hat{\Phi}'_i$ of the form: for some $n > 1$, $\hat{\Phi}'_i = \hat{\Phi}_i$ for $i > n$ and = the Dirac measure at either $\xi_1(i)$ or $\xi_2(i)$ for $i \leq n$, we have $\nu_E[\Phi'] \in Z$. The Dirac measure at ξ in the statement of the lemma is a limit of such Φ' in $\mathbb{P}_0(L)$. Then $\nu_E\{\xi\}$ is a limit point of the corresponding $\nu_E[\Phi']$ in view of the continuity of the map $\Phi \to \nu_E[\Phi]$ (Corollary V.1.2). Since Z is closed, the proof is complete. QED

Consider Z as a union of two closed line segments Z_1 and Z_2, Z_1 being the line segment between $\nu_E\{\xi_1\}$ and $\nu_E[\Phi]$, and Z_2 that between $\nu_E\{\xi_2\}$ and $\nu_E[\Phi]$. Let ξ'_n,

$n = 0,1,2,\ldots$, be a sequence in L defined as follows: $\xi'_0 = \xi_1$, $\xi'_n = [\xi_2(1),\ldots,\xi_2(n), \xi_1(n+1), \xi_1(n+2),\ldots]$, $n \geq 1$. By the above lemma, $\nu_E\{\xi'_n\} \in Z$, $n \geq 0$. Since $\xi'_n \to \xi_2$ as $n \to \infty$, we have $\nu_E\{\xi'_n\} \to \nu_E\{\xi_2\}$ by the continuity of the map $\Phi \to \nu_E[\Phi]$. Thus the sequence $\nu_E\{\xi_n\}$, $n \geq 0$, starts in Z_1 and eventually moves into Z_2. Let n be the first time this happens. Then either $\nu_E\{\xi'_n\} = \nu_E[\Phi]$, or $\nu_E[\Phi]$ is a convex combination of $\nu_E\{\xi'_n\}$ and $\nu_E\{\xi'_{n-1}\}$. Since ξ'_n and ξ'_{n-1} differ from each other only in the n-th component, the argument of Lemma V.1.2 shows that we may take $\hat{\Phi}_i$ = the Dirac measure at $\xi'_n(i)$ for $i \neq n$ and $\hat{\Phi}_n$ = a suitable convex combination of the Dirac measures at $\xi_1(n)$ and $\xi_2(n)$. In view of Lemma V.1.2, we have proved the following.

Theorem 1.1. Each extreme point of H corresponds to a $\nu_E[\Phi]$, $\Phi = \prod_i \hat{\Phi}_i$, satisfying the following. For all but at most one i, $\hat{\Phi}_i$ is a Dirac measure at some $\xi(i) \in D$ such that $p(i,.,\xi(i))$ is an extreme point of the set $\{\int p(i,.,u)\,\mu(du), \mu \in \mathbb{P}(D)\} \subset \mathbb{P}(S)$. For the single remaining i, if any, it is a convex combination of two such Dirac measures.

Next, note that the compactness of G in $\mathbb{P}(S \times D)$, and hence of G′, already guarantees the existence of an optimal SSRS in view of the lower semicontinuity of the map $\nu \in \mathbb{P}(S \times D) \to \int k\,d\nu \in R$.

Theorem 1.2. (a) The constrained problem has an optimal SSRS $\gamma[\Phi]$, $\Phi = \prod_i \hat{\Phi}_i$, satisfying the following. For all but at most m values of i, $\hat{\Phi}_i$ is a Dirac measure as in the statement of Theorem 1.1. For the remaining i, it is a convex combination of m_i such Dirac measures where the integers $\{m_i\}$ satisfy: $m_i \geq 2$ and $\sum_i m_i \leq m$.

(b) If in addition G′ has a nonempty interior, then there exist λ_i, $\psi_i \geq 0$, $1 \leq i \leq m$, such that the functional

$$\nu \to \int k d\nu - \sum_{i=1}^{m} \psi_i(\int k_i\, d\nu - \beta_i) - \sum_{i=1}^{m} \lambda_i(\alpha_i - \int k_i\, d\nu) \tag{1.3}$$

attains a minimum at the above $\gamma[\Phi]$. Also, for $\nu = \nu_E[\Phi]$,

$$\sum_{i=1}^{m} (\lambda_i(\alpha_i - \int k_i\, d\nu) + \psi_i(\int k_i\, d\nu - \beta_i)) = 0.$$

If $F(\nu, \{\lambda_i\}, \{\psi_i\})$ denotes the right hand side of (1.3), then the following 'saddle point' property holds. For all $\bar{\lambda}_i$, $\bar{\psi}_i \geq 0$, $1 \leq i \leq m$, and SSRS $\gamma[\varphi]$,

$$F\{\nu_E[\Phi], \{\bar{\lambda}_i\}, \{\bar{\psi}_i\}\} \leq F(\nu_E[\Phi], \{\lambda_i\}, \{\psi_i\})$$

$$\leq F(\nu_E[\varphi], \{\lambda_i\}, \{\psi_i\}).$$

Proof. Using Choquet's theorem as in the proof of Theorem III.1.4(a), one sees that the optimum will occur at an extreme point of G'. In turn, it is not difficult to see that the same will also be an extreme point of a set of the type

$$\{\nu \mid \int k_{i_n} d\nu \leq \text{ (or } \geq) \; \delta_{i_n}, \; 1 \leq n \leq \ell\}$$

for some $\{i_1,...,i_\ell\} \subset \{1,...,m\}$, $\ell \leq m$ and $\{\delta_1,...,\delta_\ell\} \subset \{\alpha_i, \beta_i, 1 \leq i \leq m\}$. For $\ell = 1$, (a) above reduces to Theorem 1.1. For $\ell = 2$, replace G by

$$G \cap \{\nu \mid \int k_{i_1} d\nu \leq \delta_{i_1}\} \text{ (or } \geq \text{ as the case may be)}$$

and repeat the argument to conclude. Iterating the argument, the claim follows. Part (b) is immediate from standard Lagrange multiplier theory (pp. 216–219 of [Luen]). QED

The following example shows that one cannot in general hope to improve upon the first claim above.

Example 1. Let $S = \{1,2\}$, $D = \{1,0\}$, $p(2,1,0) = p(2,1,1) = 1$, $p(1,2,0) = p_0 = 1-p(1,1,0)$, $p(1,2,1) = p_1 = 1-p(1,1,1)$ for $0 < p_0 < p_1 < 1$. Let $k(1,1) = k(1,0) = 1$, $k(2,1) = k(2,0) = 0$. The invariant probability measure under any SS that picks the control 0 when at state 1 is seen to be $[(1+p_0)^{-1}, p_0(1+p_0)^{-1}]$, leading to a cost of $(1+p_0)^{-1}$. Similarly the invariant probability measure under any SS that picks the control 1 when at state 1 is $[(1+p_1)^{-1}, p_1(1+p_1)^{-1}]$ leading to a lower cost of $(1+p_1)^{-1}$. Now, let $k_1(1,1) = k_1(1,0) = 0$ and $k_1(2,1) = k_1(2,0) = 1$. Then the average of k_1 under the two invariant probability measures above is respectively $a_0 = p_0(1+p_0)^{-1}$ and $a_1 = p_1(1+p_1)^{-1}$. Clearly, $a_1 > a_0$. Let $a = (a_0 + a_1)/2$. Consider the constraint $\int k_1 d\nu_E[\Phi] \leq a$. It is not difficult to see that under any SRS, the invariant probability measure is $[(1+p)^{-1}, p(1+p)^{-1}]$ for some $p \in [p_0, p_1]$ and the corresponding averages of k and k_1 are $(1+p)^{-1}$ and $p(1+p)^{-1}$ respectively. As p increases from p_0 to p_1, the former decreases and the latter increases. The optimum for the constrained problem is seen to occur at the value of p for which $p(1+p)^{-1} = a$. Clearly, no optimal SS exists, only an SRS.

The corresponding developments for the other cost criteria, namely (C1)-(C3), proceed along similar lines with the corresponding occupation measures replacing $\nu_E[\Phi]$. The details are omitted. However, one qualification must be made here. For ergodic control, the occupation measures and hence the foregoing developments do not depend on the initial condition. For other cost criteria, the occupation measure does depend on the initial condition and therefore the foregoing theory can be adapted only for a given initial condition.

This dependence on the initial condition cannot in general be avoided, in contrast to the unconstrained case. The following extreme example makes this point.

Example 2. Consider the discounted cost problem with $\beta = 0.01$, $p(1,2,.) \equiv p(1,3,.) \equiv 1/2$, k bounded and a single constraint $\int k_1 \, d\nu_{D1} \le 2$, with $k_1(1,.) \equiv 0$, $k_1(2,.) \equiv 100$ and $|k_i(i,.)| \le 2$ for $i \ge 3$. The set of admissible ν_{D1} then is the set of occupation measures under all SRS, while the set of admissible ν_{D2} is empty.

VII.2 Other Multiobjective Problems

This section considers some other formulations of the multiobjective control problem. For sake of being specific, we consider the ergodic control set-up with compact G. It will be clear that analogous developments are possible for other cost criteria as well.

For some $m \ge 2$, consider the vector cost criterion $[\int k_1 \, d\nu_E[\Phi], \dots, \int k_m \, d\nu_E[\Phi]]$ where $k_i \in C_b(S \times D)$, $1 \le i \le m$. In general there need not exist a $\gamma[\Phi]$ that minimizes all of $\int k_i \, d\nu_E[\Phi]$ over G. This leads to the concept of Pareto optimality. An SSRS $\gamma[\Phi]$ is said to be Pareto optimal if there does not exist any SSRS $\gamma[\varphi]$ for which

$$\int k_i \, d\nu_E[\varphi] \le \int k_i \, d\nu_E[\Phi], \quad 1 \le i \le m,$$

with the inequality being strict for at least one i. Pareto optimality is clearly the minimal requirement for any reasonable notion of an optimal solution for the multiobjective problem.

Theorem 2.1. Any SSRS (SSS) which minimizes $\sum_{i=1}^{m} \lambda_i \int k_i \, d\nu_E[\Phi]$ for some $\lambda_i > 0$, $1 \le i \le m$, is Pareto optimal. Conversely, any Pareto optimal SSRS $\gamma[\Phi]$ minimizes the above functional for some choice of $\lambda_i \ge 0$, $1 \le i \le m$.

Proof. The first claim is easy. The second is an immediate consequence of the well-known Karlin lemma ([Kar], p. 217) applied to the closed convex set

$$A = \{ [\int k_1 \, d\nu_E[\Phi], \dots, \int k_m \, d\nu_E[\Phi]] \mid \gamma[\Phi] \text{ an SSRS} \} \subset R^m. \qquad \text{QED}$$

Note that the converse is only partial, since we have $\lambda_i \ge 0$ in place of $\lambda_i > 0$. It becomes exact if A is a polyhedron, making the said condition both necessary and sufficient for Pareto optimality ([ABB], p. 87). In view of our characterization of the extreme points of G (Lemma V.1.2), this would be the case if S, D were finite sets. In general, one has the weaker conclusion that the subset of A corresponding to the Pareto optimal $\gamma[\Phi]$ obtained by minimizing $\sum_{i=1}^{m} \lambda_i \int k_i \, d\nu_E[\Phi]$ for $\lambda_i > 0$, $1 \le i \le m$, is a dense subset of the set of all elements of A that correspond to a Pareto optimal $\gamma[\Phi]$ ([ABB], p. 87).

One often reduces a vector cost criterion as above to a scalar one by introducing a 'utility function', i.e. a continuous function $U : A \to R^+$ satisfying $U([x_1, x_2,...,x_m]) \geq U([y_1, y_2,...,y_m])$ whenever $x_i \geq y_i$ for $1 \leq i \leq m$. One then minimizes $U([\int k_1 \, d\nu_E[\Phi]..., \int k_m \, d\nu_E[\Phi]])$ over G. Any $\gamma[\Phi]$ that achieves this minimum will clearly be Pareto optimal. The function $U([x_1,...,x_m]) = \sum_{i=1}^{m} \lambda_i x_i$ is a special case of a utility function and leads to the situation of Theorem 2.1.

We shall consider two more examples of utility functions. Let

$$x_i^* = \min_G \int k_i \, d\nu_E[\Phi], \quad 1 \leq i \leq m,$$

$$x^* = [x_1^*, x_2^*,...,x_m^*].$$

Call x^* the 'ideal' or 'utopian' point ([MVRS], p. 250). $U(x) = \| x - x^* \|^2$ then defines a utility function A, as can be easily verified.

Theorem 2.2. For the above choice of U, $U([\int k_1 \, d\nu_E[\Phi],..., \int k_m \, d\nu_E[\Phi]])$ attains its minimum at a unique point $\nu_E[\varphi]$ in A characterized by

$$\sum_{i=1}^{m} (\int k_i \, d\nu_E[\Phi] - \int k_i \, d\nu_E[\varphi]) \, (x_i^* - \int k_i \, d\nu_E[\varphi] \leq 0$$

for all SSRS $\gamma[\Phi]$.

Proof. This is immediate from Theorem 1, p. 69, of [Luen]. QED

When S and D are finite sets, this problem is amenable to a coupled linear-quadratic programming algorithm [Gh].

As another example of a utility function, consider

$$U([x_1,...,x_m]) = \max_i x_i.$$

This leads to the 'min-max' strategy of multiobjective control. Once again, the theory developed so far ensures the existence of an optimal SSRS.

In either case, an optimal SSS need not exist. It is not difficult to adapt Example 1 of the preceding section to illustrate this point.

Chapter VIII Control Under Partial Observations

This chapter studies the situation when the chain $\{X_n\}$ is not directly observed by the controller. Instead, he observes a related 'observations process' $\{Y_n\}$, based on which he must choose the control. The correct 'state variable' for the controller at time n is the conditional law of X_n given the observations Y_i, $i \le n$, and controls Z_i, $i \le n$ (or some variant of it, such as the 'unnormalized conditional law' introduced below). Either is computed recursively by the corresponding 'filtering equations' derived in the first section below. The original control problem can now be recast as an equivalent problem for this new state variable. The remaining sections study these equivalent control problems for cost criteria (C1)-(C4). The first section is based on Chapter 6, [KuVa], [Whi 2] and [Bor 8]. The rest of the treatment is partly based on [Bor 8], [Whi 2] and is partly new.

VIII.1 Preliminaries

Before we begin, there are some technicalities that need to be clarified. In Chapter II, we started with control spaces D(i) dependent on the state variable $i \in S$. By suitably redefining the transition probability functions, these were replaced by a common $D = \prod_i D(i)$. This caused no loss of generality there. In the case of partial observations, it does. If the control space at each time depends on the current state and is known, that amounts to an additional observation, and should be incorporated into $\{Y_n\}$. If the control space depends on the state and is not known, one is not assured of success in implementing the chosen controls, which leads to modelling difficulties. We circumvent these problems by *assuming* that all D(i)'s are identically equal to a prescribed compact metric space D.

Another technical convenience exploited so far is to pretend that at time n, the entire vector ξ_n of a CS $\{\xi_m\}$ is picked and its X_n-th component implemented as the control. This formalism runs into the same kind of difficulty as above. Hence in this chapter, we consider the D-valued process $\{Z_n\}$, $Z_n = \xi_n(X_n)$, as the prescribed control sequence instead of the L-valued process $\{\xi_n\}$. By abuse of terminology, we refer to $\{Z_n\}$ as a CS. The definitions of MRS, SRS, MS, SS will also be modified later.

Let H be a finite or countable set called the 'observation space'. Let $\{X_n\}$ be a controlled Markov chain governed by a CS $\{Z_n\}$ as above and observed via the H-valued observation process $\{Y_n\}$, the exact mechanism of which is as follows. Let $p : S \times D \times S \times H \to [0,1]$ be a family of continuous maps satisfying

$$\sum_{j,k} p(i, u, j, k) = 1, \quad i \in S, \ u \in D.$$

Let $\bar{\mathcal{F}}_n = \sigma(X_m, Z_m, Y_m, m \le n)$, $\mathcal{F}_n = \sigma(Z_m, Y_m, m \le n)$, $n \ge 0$, with X_0, Y_0 being prescribed random variables. The dynamics of $\{(X_n, Y_n)\}$ is described by

$$P(X_{n+1} = j, \ Y_{n+1} = k/\bar{\mathcal{F}}_n) = p(X_n, Z_n, j, k), \ n \ge 0, \ j \in S, \ k \in H. \tag{1.1}$$

The problem of control under partial observations is to choose $\{Z_n\}$ to minimize a given cost functional under the constraint: Z_n is conditionally independent of X_i, $i \le n$, given $\mathcal{F}_n$ for $n \ge 0$.

Introduce matrices $P(k, u) = [\,[p(i, u, j, k)\,]\,]_{i, j \in S}$, indexed by $k \in H$, $u \in D$. For $n \ge 0$, let $\pi_n(i)$ denote a version of the conditional law of X_n given Y_m, $m \le n$ and Z_m, $m < n$, which is the same as the conditional law of X_n given $\mathcal{F}_n$ in view of our condition on Z_n. Then for $i \in S$, $\pi_n(i) = P(X_n = i/\mathcal{F}_n)$ is simply the ratio of the probability of all paths consistent with observed $\{(Y_m, Z_m), \ m \le n\}$ leading to i after n time steps, to the probability of all paths consistent with observed $\{(Y_m, Z_m), \ m \le n\}$. In other words,

$$\pi_n(i) = \pi_0 \Big(\prod_{m=0}^{n-1} P(Y_{m+1}, Z_m) \Big) e_i / \pi_0 \Big(\prod_{m=0}^{n-1} P(Y_{m+1}, Z_m) \Big) 1_c, \tag{1.2}$$

where e_i = the column vector with a 1 in the i'th place and zeroes elsewhere. Let $R : \ell_1 \to \ell_1$ be the map that maps $x = [x_1, x_2, \ldots] \in \ell_1$ to $[x_1/\|x\|_1, x_2/\|x\|_1, \ldots]$ when $\|x\|_1 > 0$ and to $[0,0,\ldots]$ otherwise. Then (1.2) leads to the recursive 'filtering equation'

$$\pi_{n+1} = R(\pi_n P(Y_{n+1}, Z_n)), \ n \ge 0. \tag{1.3}$$

For $k \in H$,

$$P(Y_{n+1} = k/\mathcal{F}_n) = \pi_n P(k, Z_n) 1_c. \tag{1.4}$$

$\{\pi_n\}$ is uniquely specified by (1.3) once π_0 and $\{(Y_n, Z_n)\}$ are known, as can be seen by iterating (1.3) backwards in time. (1.3) and (1.4) together show that $\{\pi_n\}$ is a $\mathbb{P}(S)$-valued controlled Markov chain governed by the CS $\{Z_n\}$. Write its transition probability as $\bar{p}(\mu, u, d\nu) \in \mathbb{P}(\mathbb{P}(S))$ for $\mu \in \mathbb{P}(S)$, $u \in D$. From (1.4) it follows that for $f \in C_b(\mathbb{P}(S))$,

$$\int f(\nu)\bar{p}(\mu, u, d\nu) = \sum_{k \in H} (\mu P(k, u) 1_c) f(R(\mu P(k, u))).$$

Thus the map $(\mu, u) \in \mathbb{P}(S) \times D \to \bar{p}(\mu, u,.) \in \mathbb{P}(\mathbb{P}(S))$ is continuous.

In terms of $\{\pi_n\}$, costs (C1), (C3) respectively become

$$(C1') \quad E\Big[\sum_{n=0}^{\infty} \beta^n \int k(.,Z_n)d\pi_n\Big], \tag{1.5}$$

$$(C3') \quad E\Big[\sum_{n=0}^{N-1} \int \iota(n,.,Z_n)d\pi_n + \int h d\pi_N\Big], \tag{1.6}$$

(C2) is special and will be treated separately. (C4) cannot in general be written as an equivalent cost in terms of $\{\pi_n\}$, but it is natural to propose

$$(C4') \quad \limsup_{n \to \infty} \frac{1}{n} \sum_{m=0}^{n-1} \int k(.,Z_m)d\pi_m \tag{1.7}$$

as a substitute.

The evolution (1.3) involved a nontrivial nonlinear operation R, and the conditional law of Y_{n+1} given $\mathcal{F}_n$ needs an additional computation (1.4). Motivated by this, we propose an alternative state variable below, called the unnormalized conditional law.

Before we proceed, note that in the foregoing, the exact law of Y_0 does not matter. It is only the conditional law of X_0 given Y_0 that is relevant, because we can always set the law of $X_0 = \pi_0$ and require X_0, $[Z_n]$ to be independent of Y_0 without really affecting the problem. Thus we are free to change the law of Y_0 and set it equal to a prescribed $\eta \in \mathbb{P}(H)$ satisfying support $(\eta) = H$.

Consider a new probability space $(\bar{\Omega}, \bar{\mathcal{F}}, \bar{P})$ defined by $\bar{\Omega} = H^\infty$, where $\bar{\mathcal{F}}$ = the product σ-field and $\bar{P} = \eta^\infty$. Let $w = (w_0, w_1, w_2, ...)$ denote typical element of $^-$. Let $\bar{\Omega}$ $\{Y_n\}$ be the canonical process on $(\bar{\Omega}, \bar{\mathcal{F}}, \bar{P})$ defined by

$$Y_n(w) = w_n, \quad n \geq 0.$$

Later on we shall identify this with our observation process via a measure transformation. $\{Y_n]\}$ are i.i.d. with law η. By augmenting the above probability space if necessary, we construct on it D-valued random variables $\{Z_n, n \geq 0\}$ such that the conditional law of Z_n given Y_m, $m \leq n$, and Z_m, $m < n$, is the same as in our original set-up. Define $\{\mathcal{F}_n\}$ as before. Without any loss of generality, we may assume that $\bar{\mathcal{F}} = V_n \mathcal{F}_n$. Then we define an M(S)-valued process $\{\nu_n\}$ recursively by

$$\nu_{n+1} = \nu_n \bar{P}(Y_{n+1}, Z_n), \quad n \geq 0, \tag{1.8}$$

where $\nu_0 = \pi_0$ and

$$\bar{P}(Y_{n+1}, Z_n) = \eta(Y_{n+1})^{-1} P(Y_{n+1}, Z_n), \ n \geq 0.$$

Anticipating the identification of the new and the old $\{(Y_n, Z_n)\}$, we have $\pi_n = R(\nu_n)$, $n \geq 0$. Let $\Lambda_n = \nu_n 1_c$, $n \geq 0$. Then $\Lambda_0 = 1$ and $\{\Lambda_n\}$ is a nonnegative $\{\mathcal{F}_n\}$-adapted process. Let $\bar{E}[.]$ denote the expectation under $\bar{P}$.

Lemma 1.1. $(\Lambda_n, \mathcal{F}_n)$ is a martingale with $\bar{E}[\Lambda_n] = 1$ for all n.

Proof. $\bar{E}[\nu_{n+1} 1_c/\mathcal{F}_n] = \bar{E}[\eta(Y_{n+1})^{-1} \nu_n P(Y_{n+1}, Z_n)1_c/\mathcal{F}_n]$

$$= \sum_{k \in H} \eta(k)\, \eta(k)^{-1}\, \nu_n P(k, Z_n)\, 1c$$

$$= \nu_n 1_c, \ n \geq 0.$$

This proves the first claim. The second claim follows immediately. QED

Note that the martingale property would hold even if $\nu_0 \notin \mathbb{P}(S)$, the only difference being that $\bar{E}[\Lambda_n]$ need no longer be 1.

Lemma 1.1 allows us to define a new probability measure $\tilde{P}$ on $(\bar{\Omega}, \bar{\mathcal{F}})$ by

$$\frac{d\tilde{P}_n}{d\bar{P}_n} = \Lambda_n, \ n \geq 0$$

where $\tilde{P}_n$, $\bar{P}_n$ are the restrictions of $\tilde{P}, \bar{P}$ respectively to $\mathcal{F}_n$, $n \geq 0$. Let $\tilde{E}[.]$ denote the expectation under $\tilde{P}$.

Lemma 1.2. Under $\tilde{P}$, the law of $\{(Y_n, Z_n), \ n \geq 0\}$ coincides with that under our original probability measure P governing (1.1).

Proof. For $k \in H$,

$$\tilde{P}(Y_{n+1} = k/\mathcal{F}_n) = \bar{E}\,[\,I\{Y_{n+1} = k\}\, \Lambda_{n+1}/\mathcal{F}_n]/\bar{E}\,[\Lambda_{n+1}/\mathcal{F}_n]$$

$$= (\nu_n 1_c)^{-1}\, \bar{E}\,[\,I\{Y_{n+1} = k\}\, (\nu_n \bar{P}(Y_{n+1}, Z_n)1_c)/\mathcal{F}_n]$$

$$= (\nu_n 1_c)^{-1} (\nu_n \bar{P}(k, Z_n) 1_c) \bar{P}(Y_{n+1} = k / \mathcal{F}_n)$$

$$= (\nu_n 1_c)^{-1} (\nu_n P(k, Z_n) 1_c) \; \eta(k)^{-1} \eta(k)$$

$$= \pi_n P(k, Z_n) 1_c$$

$$= P(Y_{n+1} = k / \mathcal{F}_n), \;\; n \geq 0.$$

Also, for A Borel in D (with $\mathcal{F}_{-1}$ = the trivial σ-field),

$$\tilde{P}(Z_n \in A / \mathcal{F}_{n-1}, Y_n) = \bar{E}[\, I\{Z_n \in A\} \, \Lambda_n / \mathcal{F}_{n-1}, Y_n] \,/\, \bar{E}[\Lambda_n / \mathcal{F}_{n-1}, Y_n]$$

$$= \Lambda_n^{-1} \Lambda_n \bar{E}[\, I\{Z_n \in A\} / \mathcal{F}_{n-1}, Y_n]$$

$$= \bar{P}(Z_n \in A / \mathcal{F}_{n-1}, Y_n), \;\; n \geq 0.$$

The law of Y_0 is identical under both $\tilde{P}$ and P, namely η. The claim follows from this and the foregoing by induction. QED

Corollary 1.1. The joint law of $\{(Y_n, Z_n, \pi_n), \; n \geq 0\}$ is the same under $\tilde{P}$ and P.

Proof. This is immediate in view of (1.3) and the foregoing. QED

Thus the new $\{Y_n\}$, $\{Z_n\}$ are identified with the original observation and control processes via the measure transformation $\bar{P} \to \tilde{P} = P$.

Lemma 1.3. For any $f \in C_b(S \times D)$,

$$E[\int f(., Z_n) \, d\pi_n] = \bar{E}[\int f(., Z_n) \, d\nu_n], \;\; n \geq 0.$$

Proof. Both sides equal

$$\bar{E}[\, (\int f(., Z_n) \, d\pi_n) (\nu_n 1_c) \,].$$ QED

We may now consider $\{\nu_n\}$ given recursively by (1.8) as an M(S)-valued controlled Markov chain under $\bar{P}$.

The simpler evolution of $\{\nu_n\}$ is one of the advantages it has over $\{\pi_n\}$. A bigger advantage is the simpler conditional statistics of Y_{n+1} given $\mathcal{F}_n$, which offers computational benefits in the handling of the dynamic programming equations that we derive in the next section. Costs (C1), (C2) now become

$$\text{(C1}'') \quad \bar{E}\Big[\sum_{m=0}^{\infty} \beta^m \int k(.,Z_m)d\nu_m\Big], \tag{1.9}$$

$$\text{(C2}'') \quad \bar{E}\Big[\sum_{m=0}^{N-1} \int \iota(m,.,Z_m)d\nu_m + \int h d\nu_N\Big] \tag{1.10}$$

respectively, in view of Lemma 1.3. A natural replacement for (C4) is

$$\text{(C4}'') \quad \limsup_{n\to\infty} \frac{1}{n} \sum_{m=0}^{n-1} \int k(.,Z_m)d\nu_m. \tag{1.11}$$

Now consider the special case of (C2). By relabelling S if necessary, let the finite set A featuring in the definition of (C2) be given by $\{1,2,\dots,N\}$. Let $Q = \{\nu \in M(A) \mid \nu(A) \le 1\}$. Define the Q-valued process $\{\pi'_n\}$ by

$$\pi'_n(i) = E[I\{\tau > n,\ X_n = i\}/\mathcal{F}_n],\quad i \in A,\ n \ge 0.$$

Using arguments similar to those used to derive (1.2), we see that this is given by

$$\pi'_n(i) = \pi_0\Big(\prod_{m=0}^{n-1} P'(Y_{m+1}, Z_m)\Big)e_i / \hat{\pi}_0\Big(\prod_{m=0}^{n-1} P(Y_{m+1}, Z_m)\Big)1_c, \tag{1.12}$$

where

(i) $\pi_0 \in \mathbb{P}(A)$, $\hat{\pi}_0(j) = \pi_0(j)$ if $j \in A$, $= 0$ otherwise,

(ii) $e_i = [0,\dots,0,1,0,\dots,0]^T$ is an N-vector with 1 in i-th place,

(iii) $P'(Y_{m+1}, Z_m) =$ the $N \times N$ submatrix of $P(Y_{m+1}, Z_m)$ obtained by taking the rows and columns corresponding to A, appropriately ordered.

(1.12) can be rewritten as

$$\pi'_{n+1} = \pi'_n P'(Y_{n+1}, Z_n)/\pi_n P(Y_{n+1}, Z_n)1_c. \tag{1.13}$$

Clearly, $\{\pi'_n\}$ does not have a Markovian evolution, but the pair $\{(\pi'_n, \pi_n)\}$ does. The latter may be considered as our new state process. (C2) is seen to equal

$$\text{(C2}') \qquad E\Big[\sum_{n=0}^{\infty} \int \bar{k}(.,Z_n)d\pi'_n\Big] \tag{1.14}$$

for $\bar{k}$ defined by (II.2.12).

Now define an M(A)-valued process $\{\nu'_n\}$ by

$$\nu'_{n+1} = \nu'_n \bar{P}'(Y_{n+1}, Z_n), \ n \geq 0, \tag{1.15}$$

with $\nu'_0 = \pi_0$ and

$$\bar{P}'(Y_{n+1}, Z_n) = (\eta(Y_{n+1}))^{-1} P'(Y_{n+1}, Z_n), \ n \geq 0.$$

By (1.12), $\pi'_n = \Lambda_n^{-1} \nu'_n$, $n \geq 0$. Thus (1.14) equals

(C2″)
$$\bar{E}\left[\sum_{m=0}^{\infty} \int \bar{k}(.,Z_m) d\nu'_m\right]. \tag{1.16}$$

Thus the problem of minimizing (C2) with partial observations can be cast either as the problem of controlling the $Q \times \mathbb{P}(S)$-valued controlled Markov chain (π'_n, π_n), $n \geq 0$, under P with cost (1.14), or that of controlling the M(A)-valued controlled Markov chain $\{\nu'_n\}$ under $\bar{P}$ with cost (1.16). As in the case of $\{\pi_n\}$ and $\{\nu_n\}$ the latter has the advantage of relative simplicity. In particular, it has the advantage of lower dimensionality.

The correct redefinitions of MRS, MS, SRS, SS now suggest themselves. It is natural to call $\{Z_n\}$ an MRS if for each n, the conditional law of Z_{n+1} given $\mathcal{F}_n$ is a.s. equal to its conditional law given π_n (or ν_n or (π'_n, π_n) or ν'_n, as applicable). We shall call it an SRS if this conditional law is indepenent of n in addition to the above. If it is a.s. a Dirac measure in either case, we shall call it an MS in the former case and an SS in the latter. Thus $\{Z_n\}$ is an MRS if there exist measurable maps $\Phi_n : \mathbb{P}(S) \to \mathbb{P}(D)$ (resp. $M(S) \to \mathbb{P}(D)$, $Q \times \mathbb{P}(S) \to \mathbb{P}(D)$ or $M(A) \to \mathbb{P}(D)$) such that the conditional law of Z_n given $\mathcal{F}_n$ a.s. equals $\Phi_n(\pi_n)$ (resp., $\Phi_n(\nu_n)$, $\Phi_n(\pi'_n, \pi_n)$ or $\Phi_n(\nu'_n)$) for $n \geq 0$. If these maps are independent of n, it is an SRS. It is an MS if there exist measurable maps f_n : $\mathbb{P}(S) \to D$ (resp., $M(S) \to D$, $Q \times \mathbb{P}(S) \to D$ or $M(A) \to D$) such that $Z_n = f_n(\pi_n)$ (resp., $f_n(\nu_n)$, $f_n(\pi'_n, \pi_n)$ or $f_n(\nu'_n)$) for $n \geq 0$. It is an SS if all f_n's are identical. The correspondence of these definitions with our earlier definitions for the complete observations case is obvious.

VIII.2 The Discounted Cost Problem

As a prelude to our study of the discounted cost problem, we first establish a useful characterization of tightness for $\mathbb{P}(S)$-valued random variables.

Let $\mu(\alpha)$, $\alpha \in I$, be a family of $\mathbb{P}(S)$-valued random variables. For each $\alpha \in I$, define $\bar{\mu}(\alpha) \in \mathbb{P}(S)$ by

$$\int f d\bar{\mu}(\alpha) = E[\int f d\mu(\alpha)], \ f \in C_b(S).$$

Lemma 2.1. The laws of $\mu(\alpha)$, $\alpha \in I$, are tight in $\mathbb{P}(\mathbb{P}(S))$ if and only if $\bar{\mu}(\alpha)$, $\alpha \in I$, are tight in $\mathbb{P}(S)$.

Proof. Suppose $\{\bar{\mu}(\alpha), \alpha \in I\}$ is a tight set in $\mathbb{P}(S)$ and the laws of $\{\mu(\alpha), \alpha \in I\}$ are not tight in $\mathbb{P}(\mathbb{P}(S))$. Let $\bar{S} = S \cup \{\infty\}$ be the one point compactification of S as before. For $\alpha \in I$, define $\mu'(\alpha) \in \mathbb{P}(\bar{S})$ to be the measure that restricts to $\mu(\alpha)$ on S and assigns zero mass to ∞. Let the laws of $\{\mu(\alpha(n))\}$, $n \geq 1$, have no limit point in $\mathbb{P}(\mathbb{P}(S))$. Since $\mathbb{P}(\bar{S})$, and hence $\mathbb{P}(\mathbb{P}(\bar{S}))$, is compact, we can find a subsequence $\{n(k)\}$ of $\{n\}$ and a $\mathbb{P}(\bar{S})$-valued random variable μ' such that $\mu'(\alpha(n(k))) \to \mu'$ in law. By our choice of $\{\alpha(n)\}$, $\mu'(\{\infty\}) > 0$ on a set of strictly positive probability. Define $\bar{\mu}(\alpha(n))$, $n \geq 1$, $\bar{\mu}' \in \mathbb{P}(\bar{S})$ by

$$\int_{\bar{S}} f d\bar{\mu}(\alpha(n)) = E[\int_S f d\mu(\alpha(n))], \ n \geq 1,$$

$$\int_{\bar{S}} f d\bar{\mu}' = E[\int_{\bar{S}} f d\mu'],$$

for $f \in C(\bar{S})$. Then $\bar{\mu}(\alpha(n(k))) \to \bar{\mu}'$ in $\mathbb{P}(\bar{S})$ and $\bar{\mu}'(\{\infty\}) > 0$. This contradicts the tightness of $\{\bar{\mu}(\alpha), \alpha \in I\}$ in $\mathbb{P}(S)$, proving the 'if' part. Let $\varepsilon > 0$. Suppose $\{\mu(\alpha), \alpha \in I\}$ has tight laws. Then we can pick $K_1 \subset \mathbb{P}(S)$ compact (and hence tight) such that

$$P(\mu(\alpha) \in K_1) \geq 1 - \varepsilon/2, \ \alpha \in I.$$

Let $K \subset S$ be a compact (i.e., finite) set such that $\nu(K) > 1 - \varepsilon/2$ for all $\nu \in K_1$. Then for $\alpha \in I$,

$$\bar{\mu}(\alpha)(K) = E[\mu(\alpha)(K)]$$

$$= E[\mu(\alpha)(K) I\{\mu(\alpha) \in K_1\}] + E[\mu(\alpha)(K) I\{\mu(\alpha) \notin K_1\}]$$

$$\geq (1 - \varepsilon/2)^2 \geq 1 - \varepsilon.$$

QED

This has the following consequence for our problem:

Lemma 2.2. The laws of $\{(\pi_n, Z_n), n \geq 0\}$ with a prescribed π_0 form a compact subset of $\mathbb{P}((\mathbb{P}(S) \times D)^\infty)$ as $\{Z_n\}$ varies over all CS/MRS/SRS/MS/SS. In the first case, it is also convex.

Proof. The tightness of these laws is immediate from the above lemma and the tightness of the laws of the corresponding $\{X_n\}$, already established in Theorem II.3.2. Given a

$\mathbb{P}(S) \times D$-valued process (π_n, Z_n), $n \geq 0$, $\{\pi_n\}$ is a controlled Markov chain governed by $\{Z_n\}$ with transition probability function $\bar{p}(.,.,.)$ if and only if for all $f \in C_b(\mathbb{P}(S))$, $n \geq 1$ and $g \in C_b((\mathbb{P}(S) \times D)^{n+1})$, one has

$$E[(f(\pi_{n+1}) - \int f(\mu)\bar{p}(\pi_n, Z_n, d\mu))\, g((\pi_0, Z_0), (\pi_1, Z_1),...,(\pi_n, Z_n))] = 0. \tag{2.1}$$

(The 'only if' part is obvious. The 'if' part follows easily using a monotone class argument.) Since (2.1) is preserved under weak convergence of the processes concerned, the compactness claim for CS follows. That for MRS, SRS, MS and SS then follows as in the complete observations case. The convexity claim for CS follows as in Theorem II.3.3. QED

Define the discounted occupation measure $\nu_{D\mu} \in M(\mathbb{P}(S) \times D)$ for initial condition $\mu \in \mathbb{P}(S)$ by:

$$\int f d\nu_{D\mu} = E\Big[\sum_{m=0}^{\infty} \beta^m f(\pi_m, Z_m)/\pi_0 = \mu\Big],\ f \in C_b(\mathbb{P}(S) \times D).$$

Let $\bar{\nu}_{D\mu} \in M(\mathbb{P}(S))$ denote its image onto the first factor space. If $\{Z_n\}$ is an SRS, let $\Phi : \mathbb{P}(S) \to \mathbb{P}(D)$ be any version of the regular conditional law of Z_n given π_n. (The choice of n is irrelevant by the definition of an SRS.) We denote this SRS by $\gamma[\Phi]$. If $\Phi(\mu)$ is a Dirac measure at, say, $\xi(\mu)$ for $\mu \in \mathbb{P}(S)$, the SRS is an SS, denoted $\gamma\{\xi\}$. The stationary transition probability function under an SRS $\gamma[\Phi]$ will be denoted by $\bar{p}_\Phi(.,.)$ for the sake of brevity. The corresponding $\nu_{D\mu}$, $\bar{\nu}_{D\mu}$ will be denoted as $\nu_{D\mu}[\Phi]$, $\bar{\nu}_{D\mu}[\Phi]$ respectively ($\nu_{D\mu}\{\xi\}$, $\bar{\nu}_{D\mu}\{\xi\}$, respectively in the case $\gamma[\Phi] = \gamma\{\xi\}$). For $f \in C_b(\mathbb{P}(S))$,

$$\int f d\bar{\nu}_{D\mu}[\Phi] = E\Big[\sum_{m=0}^{\infty} \beta^n f(\pi_m)\Big]$$

$$= f(\mu) + \beta E\Big[E[\sum_{m=0}^{\infty} \beta^m f(\pi_{m+1})/\pi_1]\Big] \tag{2.2}$$

$$= f(\mu) + \beta E\Big[\sum_{m=0}^{\infty} \beta^m E[f(\pi_{m+1})/\pi_m]\Big], \tag{2.3}$$

where $\{\pi_n\}$ is governed by $\gamma[\Phi]$ with $\pi_0 = \mu$. Let $G_D = \{f : \mathbb{P}(S) \to M(\mathbb{P}(S)) \mid f$ is measurable and the total mass of $f(\mu)$ is bounded uniformly in $\mu \in \mathbb{P}(S)\}$.

Lemma 2.3. The map $\mu \to \bar{\nu}_{D\mu}[\Phi]$ is the unique solution in G_D to either of the following equations:

$$\int f d\bar{\nu}_{D\mu}[\Phi] = f(\mu) + \beta \int\int f d\bar{\nu}_{Dq}[\Phi]\bar{p}_\Phi(\mu, dq),\ f \in C_b(\mathbb{P}(S)), \tag{2.4}$$

$$\int f d\bar{\nu}_{D\mu}[\Phi] = f(\mu) + \beta \int\int f(q)\bar{p}_{\Phi}(u, dq)\bar{\nu}_{D\mu}[\Phi](du), \quad f \in C_b(\mathbb{P}(S)). \tag{2.5}$$

Proof. That $\bar{\nu}_{D\mu}[\Phi]$ satisfies (2.4), (2.5) is immediate from (2.2), (2.3). Iterating either of (2.4), (2.5), we have

$$\int f d\bar{\nu}_{D\mu}[\Phi] = f(\mu) + \sum_{m=1}^{N} \beta^m \int f(q)\bar{p}^m_{\Phi}(\mu, dq) + R_N \tag{2.6}$$

for $N \geq 1$, where

$$\bar{p}^m_{\Phi}(\mu, dq) = \underbrace{\int \ldots\ldots\ldots\ldots \int}_{(m-1)\ \text{times}} \bar{p}_{\Phi}(\mu, dq_1)\bar{p}_{\Phi}(q_1, dq_2)\ldots\bar{p}_{\Phi}(q_{m-1}, dq)$$

and R_N is a remainder term satisfying

$$|R_N| \leq K\beta^N$$

for a suitable $K > 0$. Letting $N \to \infty$ in (2.6), one gets an explicit series representation for its left hand side, proving the uniqueness claim. QED.

Lemma 2.4. For a given $\mu \in \mathbb{P}(S)$, the set of attainable $\nu_{D\mu}$ under all CS is the same as that under all SRS and is compact convex. The corresponding set under all SS is a compact subset of the above.

Proof. The first claim and the compactness claims are proved in the same way as for the complete observations case. Convexity may be read off Lemma 2.2. An alternative proof goes as follows. Given two SRS $\gamma[\Phi_1]$, $\gamma[\Phi_2]$, and an $a \in (0,1)$, define

$$\nu = a\,\bar{\nu}_{D\mu}[\Phi_1] + (1-a)\,\bar{\nu}_{D\mu}[\Phi_2] \in M(\mathbb{P}(S));$$

$$\Lambda = a d\bar{\nu}_{D\mu}[\Phi_1]/d\nu = 1 - (1-a)d\bar{\nu}_{D\mu}[\Phi_2]/d\nu \geq 0.$$

Define $\Phi' : \mathbb{P}(S) \to \mathbb{P}(D)$ by

$$\Phi'(q) = \Lambda\Phi_1(q) + (1-\Lambda)\Phi_2(q), \quad q \in \mathbb{P}(S).$$

Write (2.5) for $\Phi = \Phi_1, \Phi_2$, multiply the first by a and the second by $(1-a)$ and add. In view of the above, one obtains

$$\int f d\nu = f(\mu) + \beta \int\int f(q)\bar{p}_{\Phi'}(u, dq)\nu(du), \quad f \in C_b(\mathbb{P}(S)).$$

Thus $\nu = \bar{\nu}_{D\mu}[\Phi']$, leading to

$$\nu_{D\mu}[\Phi'] = a\,\nu_{D\mu}[\Phi_1] + (1 - a)\nu_{D\mu}[\Phi_2].$$ QED.

Corollary 2.1. An optimal SRS exists.

This is again proved exactly as in the complete observations case. We do not prove that the extreme points of the above convex set correspond to an SS. Nor do we claim that an optimal SS exists. The convex analytic arguments used to prove such results in the complete observations case do not seem to extend to the present set-up. We shall obtain these results only indirectly via the dynamic programming equation.

Define the value function $V : \mathbb{P}(S) \to R^+$ by

$$V(\mu) = \min_{CS} E\Big[\sum_{m=0}^{\infty} \beta^m \int k(.,Z_m)d\pi_m / \pi_0 = \mu\Big]. \tag{2.7}$$

Lemma 2.4. V is continuous.

Proof. Given $\mu \in \mathbb{P}(S)$ and a D-valued random sequence $\{Z_n\}$ on a probability space, we may augment the probability space if necessary (e.g., by attaching to it a copy of $H^\infty \times \mathbb{P}(S)^\infty$) and construct on it random processes $\{\pi_n\}$, $\{Y_n\}$ that evolve as in (1.3) and (1.4) with $\pi_0 = \mu$. Thus it makes sense to denote the expectation on the right hand side of (2.7) as $F(\{Z_n\}, \mu)$. Then

$$V(\mu) = \min_{\{Z_n\}} F(\{Z_n\}, \mu).$$

From (1.3) and (1.4), it is clear that for a prescribed $\{Z_n\}$ as above, the law of the resultant $\{\pi_n\}$ will vary continuously with μ. Let $\mu_n \to \mu_\infty$ in $\mathbb{P}(S)$. For n = 1,2,..., let $\{Z^n_m\}$ be an optimal control sequence for initial condition μ_n. Let $\{\pi^n_m\}$ denote the corresponding processes of conditional laws. By dropping to a subsequence if necessary and invoking Lemma 2.2, we may assume that $\{\pi^n_m\}$, $\{Z^n_m\}$ jointly converge in law to $\{\pi^\infty_m\}$, $\{Z^\infty_m\}$ where $\{\pi^\infty_m\}$ is a $\mathbb{P}(S)$-valued controlled Markov chain evolving as in (1.3)-(1.4) under the control sequence $\{Z^\infty_m\}$. Then for any control sequence $\{Z_m\}$ on some probability space,

$$F(\{Z_m\}, \mu_\infty) \leftarrow F(\{Z_m\}, \mu_n) \geq V(\mu_n) = F\{Z^n_m\}, \mu_n) \to F(\{Z^\infty_m\}, \mu_\infty).$$

Since $\{Z_m\}$ was arbitrary, $F(\{Z^\infty_m\}, \mu_\infty) = V(\mu_\infty)$, proving the claim. QED.

Lemma 2.6. V satisfies

$$V(\mu) = \min_{u \in \mathbb{P}(D)} \left(\int\int k(x,y)\mu(dx)u(dy) + \beta\int\int u(dy)\bar{p}(\mu, y, d\nu)V(\nu)\right) \tag{2.8}$$

$$= \min \left(\int\int k(x,y)\Phi(\mu)(dy)\mu(dx) + \beta\int \bar{p}_{\Phi}(\mu, d\nu)V(\nu)\right) \tag{2.9}$$

where the minimum in (2.9) is over all measurable maps $\Phi : \mathbb{P}(S) \to \mathbb{P}(D)$.

The map

$$q \in \mathbb{P}(D) \to \int\int k(x,y)q(dy)\mu(dx) + \beta\int\int \bar{p}(\mu,y,d\nu)V(\nu)q(dy) \tag{2.10}$$

is continuous in q for each fixed $\mu \in \mathbb{P}(S)$. Thus the minimum over q is always attained for each fixed μ. In fact, it will be attained on the set of Dirac measures on D. The selection theorem of Lemma 1, [Ben], then allows us to pick a measurable map $\Phi : \mathbb{P}(S) \to \mathbb{P}(D)$ such that $\Phi(\mu)$ attains the minimum in (2.10) at each μ. Thus the right hand side of (2.9) is well-defined. Similar remarks will apply to analogous situations that arise later in the chapter.

The lemma is proved along the lines of the corresponding result in the complete observations case. The same applies to its natural sequel, Lemma 2.7 below.

Lemma 2.7. An SRS $\gamma[\Phi]$ is optimal for any initial condition if and only if $\gamma[\Phi]$ attains the minimum in (2.9) for each μ.

As already noted, the minimum is attained by a measurable $\Phi : \mathbb{P}(S) \to \mathbb{P}(D)$ taking values in the set of Dirac measures on D and hence identifiable with a measurable $\xi : \mathbb{P}(S) \to D$ such that $\Phi(\mu)$ is supported at $\xi(\mu)$ for all $\mu \in \mathbb{P}(S)$. Thus the above can be improved to the following:

Theorem 2.1. V satisfies

$$V(\mu) = \min_{u \in D} \left(\int k(x,u)\mu(dx) + \beta\int \bar{p}(\mu, u, d\nu)V(\nu)\right) \tag{2.11}$$

$$= \min \left(\int k(x, \xi(\mu))\mu(dx) + \beta\int p(\mu, \xi(\mu), d\nu)V(\nu)\right) \tag{2.12}$$

where the minimum in (2.12) is over all measurable $\xi : \mathbb{P}(S) \to D$. Both minima are attained. An SS $\gamma\{\xi\}$ is optimal for any initial condition if and only if ξ attains the above minimum.

Note an important difference with the complete observations case. We do not claim that an SRS or an SS optimal for one initial condition will necessarily be so for all initial conditions. What fails in extending the corresponding argument from the complete observations case is that the chain $\{\pi_n\}$ need not have a 'single communicating class' under all SRS; i.e. $\mathbb{P}(S)$ may decompose into two or more disjoint non-empty sets each separately closed under the evolution (1.3)-(1.4).

The following two results are proved by arguments analogous to those for the corresponding results in the complete observations case.

Theorem 2.2. V is the (pointwise) minimal non-negative solution to (2.8) or (2.11). It is the unique bounded solution if k is bounded.

Let $V\{\xi\}(\mu)$, $\mu \in \mathbb{P}(S)$, stand for

$$E\Big[\sum_{m=0}^{\infty} \beta^m \int k(.,\xi(\pi_m))d\pi_m / \pi_0 = \mu\Big]$$

with $\{\pi_m\}$ governed by $\gamma\{\xi\}$.

Theorem 2.3. $\gamma\{\xi\}$ is optimal for any initial condition if and only if

$$V\{\xi\}(\mu) = \min_{\xi} (\int k(x, \xi(\mu))\mu(dx) + \int \bar{p}(\mu, \xi(\mu), d\nu)V\{\xi\}(\nu)). \tag{2.13}$$

Now we shall translate the above results into the language of $\{\nu_n\}$, $\bar{P}$. For a measurable $\xi : \mathbb{P}(S) \to D$, define $\xi' : M(S) \to D$, by $\xi'(\mu) = \xi(\bar{\mu})$ where $\bar{\mu} = \mu(S)^{-1}\mu$ when $\mu(S) > 0$ and any arbitrary element of $\mathbb{P}(S)$ when $\mu(S) = 0$. Since the evolution of $\{\nu_n\}$ is linear with respect to the initial condition, the cost under any SS $\gamma\{\varphi\}$ when $\nu_0 = \mu$ is $\mu(S)$ times the cost when $\nu_0 = \bar{\mu}$. It follows that if $\gamma\{\xi\}$ is as in Theorem 2.3, $\gamma\{\xi'\}$ for ξ' defined as above is optimal for the controlled Markov chain $\{\nu_n\}$ for any initial condition. From (2.11), we have

$$V(\mu) = \min_{CS} E[\int k(.,Z_0)d\pi_0 + \beta V(\pi_1)/\pi_0 = \mu], \ \mu \in \mathbb{P}(S).$$

Extend V to M(S) by writing $V(\mu) = \mu(S)V(\bar{\mu})$. Let $\{\nu_n\}$ evolve as in (1.8) with $\nu_0 = \mu \in M(S)$ and with $\bar{P}$ as the underlying probability measure. Then

$$\begin{aligned} V(\mu) &= \mu(S)V(\bar{\mu}) \\ &= \mu(S) \min_{CS} E[\int k(.,Z_0)d\pi_0 + \beta V(\pi_1)/\pi_0 = \bar{\mu}] \\ &= \mu(S) \min_{CS} E[\int k(1,Z_0)d\nu_0 + \beta(\nu_1 1_c)^{-1} V(\nu_1)/\nu_0 = \bar{\mu}] \\ &= \mu(S) \min_{CS} \bar{E}[\int k(\cdot,Z_0)d\nu_0 + \beta V(\nu_1)/\nu_0 = \bar{\mu}]. \end{aligned}$$

$$= \min_{CS} \bar{E}[\int k(.,Z_0)d\nu_0 + \beta V(\nu_1)/\nu_0 = \mu]. \tag{2.14}$$

A similar argument leads to

$$V(\mu) = \min_{CS} \bar{E}\Big[\sum_{m=0}^{\infty} \beta^m \int k(.,Z_m)d\nu_m/\nu_0 = \mu\Big],$$

as desired. Let the transition probability function for $\{\nu_n\}$ be denoted by $q(\mu, u, d\nu)$.

Theorem 2.4. $V : M(S) \to R^+$ defined as above satisfies

$$V(\mu) = \min_u (\int k(x, u)\mu(dx) + \beta \int q(\mu, u, d\nu)V(\nu)) \tag{2.15}$$

$$= \min_\xi (\int k(x, \xi(\mu))\mu(dx) + \beta \int q(\mu, \xi(\mu), d\nu)V(\nu)), \tag{2.16}$$

where the minimum in (2.16) is over all measurable $\xi : M(S) \to D$. Both minima are attained and an SS $\gamma\{\xi\}$ is optimal for any initial condition if and only if $\xi(\mu)$ attains the minimum in (2.15) for each μ.

The proof is straightforward in view of (2.14) above. Also, the obvious analogue of Theorem 2.3 holds.

In conclusion, we note that the dynamic programming equations (2.15) - (2.16) are not as formidable as they seem. In view of the easily verifiable fact $V(\alpha\mu) = \alpha V(\mu)$ for $\alpha \geq 0$, a simple computation reduces (2.15) - (2.16) to

$$V(\mu) = \min_u (\sum_i \mu(i)k(i, u) + \beta \sum_k V(\mu\, P(k, u))) \tag{2.17}$$

$$= \min_\xi (\sum_i \mu(i)k(i, \xi(\mu)) + \beta \sum_k V(\mu\, P(k, \xi(\mu)))). \tag{2.18}$$

VIII.3 Finite Time Control Problems

Consider (C2). Note that $\mu_n \to \mu_\infty$ in M(A) if and only if either $\mu_n(A) \to 0 = \mu_\infty(A)$ or $\mu_\infty(A) > 0$ and $\mu_n(A)^{-1}\mu_n \to \mu_\infty(A)^{-1}\mu_\infty$ in $\mathbb{P}(A)$. Since $\mathbb{P}(A)$ is compact, this implies that any subset of M(A) whose elements have uniformly bounded total mass will be relatively compact in M(A). Thus Q is relatively compact. As Q is easily seen to be closed in M(A), it is in fact compact. By Lemma 2.2, we conclude that the law of (π'_n, π_n),

$n \geq 0$, as a $Q \times \mathbb{P}(S)$ valued process remains in a tight set as the control sequence is varied over all CS while π_0 is held fixed. An argument analogous to that of Lemma 2.2 then yields the following.

Lemma 3.1. The laws of $(\pi'_n, \pi_n)\}$, $n \geq 0$, over all CS, MRS, MS, SRS or SS form a comapct subset of $\mathbb{P}((Q \times \mathbb{P}(S))^\infty)$, the initial condition being held fixed. In case of CS, it is also convex.

From now on, we write $\hat{\pi}_n = (\pi'_n, \pi_n)$, $n \geq 0$, and consider $\{\hat{\pi}_n\}$ as a $Q \times \mathbb{P}(S)$-valued controlled Markov chain whose transition probability function is denoted by $\hat{p}(\mu,u,d\nu) \in \mathbb{P}(Q \times \mathbb{P}(S))$ for $\mu \in Q \times \mathbb{P}(S)$, $u \in D$.

We shall say that a function $f \in C(M(A) \times \mathbb{P}(\bar{S}) \times D)$ is a tame function if there exist $\bar{n} \geq 1$ and $g \in C_b(A^{\bar{n}} \times M(A) \times \mathbb{P}(\bar{S}) \times D)$ such that

$$f(\nu, \pi, u) = \int \dots \int g(x_1,\dots,x_{\bar{n}}, \nu, \pi, u)\nu(dx_1)\dots\nu(dx_{\bar{n}}). \tag{3.1}$$

Let K_T denote the space of finite linear combinations of tame functions and K_T^* the space of bounded linear functionals on it with the coarsest topology that renders continuous the maps $\nu \in K_T^* \to \nu(f) \in R$ for $f \in K_T$. Let $G_T \subset K_T$ be the subset consisting of functions that do not have an explicit dependence on the last argument, and thus can be considered as functions on $M(A) \times \mathbb{P}(\bar{S})$.

Let $f \in K_T$ be as in (3.1) and $K > 0$ a bound on $|g(.)|$. Then

$$|f(\hat{\pi}_m, Z_m)| \leq (\pi'_m(A))^{\bar{n}} K.$$

Recall from Chapter IV that

$$E[\pi'_n(A)] = P(T > n) \leq K' a^n$$

for a suitable $K' > 0$ and an $a \in (0,1)$. Thus

$$E[|f(\hat{\pi}_n, Z_n)|] \leq KK' a^n, \quad n \geq 0. \tag{3.2}$$

This allows us to define for $\mu \in Q \times \mathbb{P}(S)$ a $\nu_{T\mu} \in K_T^*$ by

$$\nu_{T\mu}(f) = E[\sum_{m=0}^{\infty} f(\hat{\pi}_m, Z_m)/\hat{\pi}_0 = \mu], \quad f \in K_T. \tag{3.3}$$

Let $\bar{\nu}_{T\mu}$ denote the restriction of $\nu_{T\mu}$ to G_T. If $\{\hat{\pi}_n\}$ is governed by an SRS $\gamma[\Phi]$, denote

$\nu_{T\mu}$, $\bar{\nu}_{T\mu}$ by $\nu_{T\mu}[\Phi]$, $\bar{\nu}_{T\mu}[\Phi]$ respectively. The transition probability under $\gamma[\Phi]$ will be denoted by $\hat{p}_{\Phi}(.,.)$. It is easy to check that for $f \in K_T$ and $h : Q \times \mathbb{P}(S) \to R$ defined by

$$h(\mu) = \nu_{T\mu}[\Phi](f), \ \mu \in Q \times \mathbb{P}(S),$$

the following hold. For $\mu = (\mu', \bar{\mu})$,

$$h(\mu) = \int\int k(x, y)\Phi(\mu)(dy)\mu'(dx) + \int p_{\Phi}(\mu, d\nu)h(\nu) \ \forall \mu, \tag{3.4}$$

and for $Q \times \mathbb{P}(S)$-valued random variables $\{\mu_n\}$ satisfying $E[\mu'_n(A)] \to 0$ with $\mu_n = (\mu'_n, \bar{\mu}_n)$,

$$| E[h(\mu_n)] | \le (C \sum_{m=0}^{\infty} a^m) E[\mu'_n(A)] \to 0 \text{ as } n \to \infty \tag{3.5}$$

for a suitable $C > 0$.

Lemma 3.2. Given any $\mu \in Q \times \mathbb{P}(S)$, a CS $\{Z_n\}$ and the corresponding $\nu_{T\mu}$, there exists an SRS $\gamma[\Phi]$ such that $\nu_{T\mu} = \nu_{T\mu}[\Phi]$.

Proof. This follows along the lines of Theorem III.1.1 in view of (3.4) and (3.5) above. Of course, one needs to 'disintegrate' $\nu_{T\mu}$. In order to do this, observe that K_T (resp. G_T) is dense in $C_b(M(A) \times \mathbb{P}(\bar{S}) \times D)$ (resp. $C_b(M(A) \times \mathbb{P}(\bar{S}))$) by the Stone-Weirstrass theorem and $\nu_{T\mu}$ (resp. $\bar{\nu}_{T\mu}$) extend uniquely to a positive Radon measure on $C_b(M(A) \times \mathbb{P}(\bar{S}) \times D)$ (resp. $C_b(M(A) \times \mathbb{P}(\bar{S}))$) with finite total mass. QED

Lemma 3.3. The set of $\nu_{T\mu}[\Phi]$ over all $\gamma[\Phi]$ is convex and sequentially compact for fixed $\mu \in Q \times \mathbb{P}(S)$.

Proof. Convexity follows from Lemmas 3.1, 3.2. For sequential compactness, it suffices, in view of Lemma 3.1, to prove the following. If $\{\hat{\pi}^n_m, m \ge 0\}$, $n = 1,2,...$, are governed by $\{Z^n_m, m \ge 0\}$, $n = 1,2,...$, respectively and $\{\hat{\pi}^n_m\}$, $\{Z^n_m\}$ converge jointly in law to some $\{\hat{\pi}^\infty_m\}$, $\{Z^\infty_m\}$ with $\{\hat{\pi}^\infty_m\}$ governed by $\{Z^\infty_m\}$, then the $\nu_{T\mu}$'s corresponding to $\{\hat{\pi}^n_m\}$, $\{Z^n_m\}$ converge to that corresponding to $\{\hat{\pi}^\infty_m\}$, $\{Z^\infty_m\}$. Using Skorohod's theorem, we may assume the former convergence to be a.s. Thus for any $f \in K_T$,

$$E[f(\hat{\pi}^n_m, Z^n_m)] \to E[f(\hat{\pi}^\infty_m, Z^\infty_m)], \ m \ge 0,$$

as $n \to \infty$. In view of (3.2), we can use the dominated convergence theorem to conclude that

$$E[\sum_{m=0}^{\infty} f(\hat{\pi}^n_m, Z^n_m)] = \sum_{m=0}^{\infty} E[f(\hat{\pi}^n_m, Z^n_m)] \to \sum_{m=0}^{\infty} E[f(\hat{\pi}^{\infty}_m, Z^{\infty}_m)] = E[\sum_{m=0}^{\infty} f(\hat{\pi}^{\infty}_m, Z^{\infty}_m)].$$

The claim follows. QED

Corollary 3.1. An optimal SRS exists.

This follows by the usual arguments. Once again we are unable to improve the above to prove the existence of an optimal SS by establishing the extremality of the $\nu_{T\mu}$ corresponding to the SS, and have to depend on the dynamic programming equatons for the purpose. Define the value function $V : Q \times \mathbb{P}(S) \to R^+$ by

$$V(\mu) = \inf_{CS} E\Big[\sum_{m=0}^{\infty} \int \bar{k}(x, Z_m) d\pi'_m / (\pi'_0, \pi_0) = \mu\Big].$$

We can now mimic the developments of the preceding section to deduce the following results.

Theorem 3.1. V satisfies

$$V(\mu) = \min_{u \in D} (\int \bar{k}(x, u)\mu'(dx) + \int \hat{p}(\mu, u, d\nu)V(\nu)) \tag{3.6}$$

$$= \min (\int \bar{k}(x, \xi(\mu))\mu'(dx) + \int \hat{p}(\mu, \xi(\mu), d\nu)V(\nu)), \tag{3.7}$$

where $\mu = (\mu, \bar{\mu}) \in Q \times \mathbb{P}(S)$, and the minimum in (3.7) is over all measurable $\xi : Q \times \mathbb{P}(S) \to D$. Both minima are attained and an SS $\gamma\{\xi\}$ is optimal for any initial condition if and only if $\xi(\mu)$ attains the minimum in (3.6) for each μ.

Once again an SS optimal for one initial condition need not be so for another. An analogue of Theorem 2.4 also holds; we omit the statement of it.

Next we translate this into the corresponding results for $\{\nu'_n\}, \bar{P}$. Define $V' : M(A) \to R^+$ by

$$V'(\mu) = \min_{CS} \bar{E}\Big[\sum_{m=0}^{\infty} \int \bar{k}(., Z_m) d\nu'_m / \nu'_0 = \mu\Big],$$

$$= V(\bar{\mu}, \hat{\mu})\mu(A)$$

where $\bar{\mu} = \mu(A)^{-1}\mu \in \mathbb{P}(A)$, and $\hat{\mu} \in \mathbb{P}(S)$ is the element that restricts to $\bar{\mu}$ on A, assigning zero mass to A^c. (This leaves out the case $\mu(A) = 0$, in which case we set $V(\mu) = 0$.) Let $\hat{q}(.,.,.)$ denote the transition probability function of $\{\nu'_n\}$. The following analogue of Theorem 2.5 can now be proved by analogous methods.

Theorem 3.2. V' satisfies

$$V'(\mu) = \min_{u \in D} \left(\int k(x, u)\mu(dx) + \int \hat{q}(\mu, u, d\nu)V'(\nu)\right) \tag{3.8}$$

$$= \min \left(\int k(x, \xi(\mu))\mu(dx) + \int \hat{q}(\mu, \xi(\mu), d\nu)V'(\nu)\right), \tag{3.9}$$

where the minimum in (3.9) is over all measurable $\xi : M(A) \to D$. Both minima are attained and an SS $\gamma\{\xi\}$ is optimal for any initial condition if and only if $\xi(\mu)$ attains the minimum in (3.9) for each μ.

Finally, one can reduce (3.8)-(3.9) to (2.17)-(2.18) with V replacing V' and $\beta = 1$ by a simple computation.

This completes our discussion of (C2). As for (C3) under partial observations, one can still convert it into a problem of control up to a first exit time using arguments analogous to those of Chapter IV. This is so because the 'exit time' therein is deterministic. It is also possible to directly adapt the methods of preceding section to derive the desired results. The final results are just what one expects and the details are straightforward. We omit both.

VIII.4 The Ergodic Control Problem

The ergodic control problem under partial observations is much harder than in the complete observations case. We shall derive some partial results here using $\{\pi_n\}$ as our state process. The corresponding definitions of SRS, SS etc. will apply.

Any $\mu \in \mathbb{P}(\mathbb{P}(S) \times D)$ can be disintegrated as

$$\mu(dq, du) = \bar{\mu}(dq)\Phi(q)(du) \tag{4.1}$$

where $\bar{\mu}$ is the image of μ under the projection $\mathbb{P}(S) \times D \to \mathbb{P}(S)$, and $\Phi : \mathbb{P}(S) \to \mathbb{P}(D)$ is the regular conditional law, defined $\bar{\mu}$ a.s. We shall always work with one arbitrary representative of this equivalence class. Define $\Gamma \subset \mathbb{P}(\mathbb{P}(S) \times D)$ by

$$\Gamma = \{\mu \in \mathbb{P}(\mathbb{P}(S) \times D) \mid \text{For } \bar{\mu},\ \Phi \text{ as in (4.1)},\ \bar{\mu} \text{ is invariant under the SRS } \gamma\{\Phi\}\}$$

$$= \{\mu \in \mathbb{P}(\mathbb{P}(S)\times D) \mid \text{For } \bar{\mu}, \Phi \text{ as in (4.1)}, \int f d\bar{\mu} = \iint f(q)\bar{p}_{\Phi}(\nu, dq)\bar{\mu}(d\nu) \text{ for all } f \in C_b(\mathbb{P}(S))\}.$$

From the second definition, one can easily check that Γ is closed.

Note that the set of invariant probability measures for the chain $\{\pi_n\}$ controlled by an SRS $\gamma[\Phi]$, when nonempty, need not be a singleton. In general it will form a closed convex set in $\mathbb{P}(\mathbb{P}(S))$, the extreme points of which correspond to ergodic measures. That is, the above chain with one of these extreme measures (say, μ) as the initial condition will be ergodic. Then (1.7) will a.s. equal $\int \hat{k}\, d\mu$ where for $\nu \in \mathbb{P}(S)$, $u \in D$,

$$\hat{k}(\nu, u) = \int k(x, u)\nu(dx). \tag{4.2}$$

In view of the ergodic decomposition of a stationary Markov chain, this will also be the case for other invariant measures (which will be convex combinations of the ergodic ones). Define

$$\beta = \inf_{\mu \in \Gamma} \int \hat{k}\, d\mu. \tag{4.3}$$

We shall assume that β is finite. We consider two alternative conditions under which the above infimum will be shown to be a minimum.

(1) (Near-monotone case.) k satisfies

$$\lim_{i \to \infty} \inf_{u} k(i, u) = \infty. \tag{4.4}$$

(2) (Stable case.) Condition A of section V.2 holds.

Observe that the 'near-monotonicity' condition here is more restrictive than the one used in the 'complete observations' case.

Lemma 4.1. Under either of the above conditions, the map $\mu \to \int \hat{k}\, d\mu$ attains its minimum on Γ.

Proof. Consider the near-monotone case to start with. Let $\{\mu_n\} \subset \Gamma$ be such that

$$\infty > \int \hat{k}\, d\mu_n \downarrow \beta. \tag{4.5}$$

Suppose $\{\mu_n\}$ has no limit point in $\mathbb{P}(\mathbb{P}(S) \times D)$. When viewed as a sequence in $\mathbb{P}(\mathbb{P}(\bar{S}) \times D)$ by identifying $\mathbb{P}(S)$ with a subset of $\mathbb{P}(\bar{S})$ in the obvious manner, it will be relatively compact. Let μ be a limit point of $\{\mu_n\}$ in $\mathbb{P}(\mathbb{P}(\bar{S}) \times D)$. Then clearly

$$\mu((\mathbb{P}(\bar{S})\backslash \mathbb{P}(S)) \times D) > 0.$$

That is, if (ξ, φ) is a $(\mathbb{P}(\bar{S}) \times D)$-valued random variable with law μ, we must have $\xi(\{\infty\}) > 0$ with strictly positive probability. Thus $\int \hat{k}\, d\mu = \infty$, contradicting (4.5) in view of

Skorohod's theorem coupled with Fatou's lemma. Hence $\mu_n \to \mu$ along a subsequence for some $\mu \in \mathbb{P}(\mathbb{P}(S) \times D)$. Since Γ is closed, $\mu \in \Gamma$. By Fatou's lemma applied as above to (4.5),

$$\beta \geq \int \hat{k}\, d\mu \geq \beta,$$

implying the claim. Now consider the stable case. For any $\mu \in \Gamma$ with decomposition (4.1), let $\{\pi_n\}$ be the stationary chain governed by $\gamma[\Phi]$ with initial law $\bar{\mu}$. Then for any $f \in C_b(S)$,

$$\begin{aligned} E[f(X_n)] &= E[\int f d\pi_n] = E[\int f d\pi_{n+1}] \\ &= E[f(X_{n+1})] \\ &= E\Big[\sum_{j \in S} p(X_n, j, Z_n) f(j)\Big] \\ &= E\Big[\sum_{i \in S} \pi_n(i) \sum_{j \in S} p(i, j, Z_n) f(j)\Big]. \end{aligned} \tag{4.6}$$

Define $\mu' \in \mathbb{P}(S \times D)$ by

$$\int f d\mu' = \int (f(x, u)\nu(dx))\mu(d\nu, du), \quad f \in C_b(S \times D).$$

Disintegrate μ' as

$$\mu'(\{i\}, du) = \tilde{\mu}(\{i\})\hat{\Phi}'_i(du), \quad i \in S.$$

Set $\Phi' = \prod_i \hat{\Phi}'_i \in \mathbb{P}_0(L)$. (4.6) then reduces to

$$\sum_i \tilde{\mu}(\{i\}) f(i) = \sum_i \sum_j f(j) p_{\Phi'}(i, j) \tilde{\mu}(\{i\}),$$

implying $\tilde{\mu} = \pi[\Phi']$. Under condition A of section V.2, the set $\{\pi[\varphi] \mid \gamma[\varphi] \text{ an SRS}\}$ is compact in $\mathbb{P}(S)$. By Lemma 2.1 and the compactness of D it follows that the set Γ itself is compact. The rest follows as in the near-monotone case. QED

Define the $\mathbb{P}(\mathbb{P}(S) \times D)$-valued process $\{\eta_n\}$ by

$$\int f d\eta_n = \frac{1}{n} \sum_{m=0}^{n-1} f(\pi_m, Z_m), \; n \geq 1, \; f \in C_b(\mathbb{P}(S) \times D),$$

where $\{\pi_m\}$ is governed by the CS $\{Z_m\}$.

Lemma 4.2. Outside a set of zero probability, any limit point of $\{\eta_n\}$ in $\mathbb{P}(\mathbb{P}(S) \times D)$ lies in Γ.

Proof. Let $\{f_m\} \subset C_b(\mathbb{P}(S) \times D)$ be a countable convergence determining class for $\mathbb{P}(\mathbb{P}(S) \times D)$. By the martingale stability theorem of [Lo], page 53, we have for $f \in C_b(\mathbb{P}(S))$,

$$\lim_{n \to \infty} \frac{1}{n} \sum_{m=0}^{n-1} (f(\pi_m) - E[f(\pi_m)/\mathcal{F}_{m-1}])$$

$$= \lim_{n \to \infty} \frac{1}{n} \sum_{m=0}^{n-1} (f(\pi_{m+1}) - \int f(\nu)\bar{p}(\pi_m, Z_m, d\nu))$$

$$= 0 \text{ a.s.} \tag{4.7}$$

Let N be the set of zero probability outside which the above holds for $f = f_m$, $m \geq 1$. Fix a sample path outside N and take any limit point μ of $\{\eta_n\}$ in $\mathbb{P}(\mathbb{P}(S) \times D)$. (There may be none.) Disintegrating μ as in (4.1), (4.7) leads to

$$\int f_m \, d\bar{\mu} = \int\int f_m(q)\bar{p}_\Phi(\nu, dq)\bar{\mu}(d\nu), \; n \geq 1.$$

The claim follows. QED

Consider the near-monotone case. Suppose that for a given sample path, a subsequence of $\{\eta_m\}$ has no limit point in $\mathbb{P}(\mathbb{P}(S) \times D)$. Arguments similar to those of the first half of the proof of Lemma 4.1 show that the cost must go to $+\infty$ along this subsequence. In view of Lemma 4.2, this leads to

$$\liminf_{n \to \infty} \frac{1}{n} \sum_{m=0}^{n-1} \int k(.,Z_m) d\pi_m \geq \beta \text{ a.s.} \tag{4.8}$$

Along with Lemma 4.1, this would seem to lead to the existence of an optimal SRS. There is, however, one catch. It is not a priori clear that any initial law for π_0 would be in the domain of attraction of the element of Γ that minimizes the cost (or, for that matter, whether this domain of attraction can be reached in a finite random time from any initial law under *some* control). Similar 'reachability' problems surface when one tries to extend the dynamic programming equations of Chapter VI to this problem. (For some results under somewhat stringent conditions, see [Plat].)

Finally, one can prove the convexity of Γ by a simple adaptation of the second proof of convexity in Lemma 2.4. Again, it is unclear how (and whether) one can characterize the extreme points of Γ as those corresponding to SS. As even the dynamic programming equations are unavailable, the existence of an optimal SS remains an open issue in general.

In the stable case, even Lemma 4.2 (and hence (4.8)) is unavailable.

Chapter IX
Adaptive Control: The Raw Self-Tuner

This chapter studies the self-tuning control of Markov chains depending on an unknown parameter with the ergodic cost criterion. The first two sections deal respectively with parameter estimation by the maximum likelihood method, and the parameter dependence of the value function. The third section describes the self-tuner and proves its optimality under a suitable 'identifiability condition'. Section 4 discusses some consequences of the lack of this condition. One scheme for dealing with these consequences will be discussed in the next chapter.

This chapter is based on [BoGh], [BoV2] and [BoV3].

X.1 Parameter Estimation

From now on, our transition probability function p will have an additional argument, namely an unknown parameter θ belonging to a compact metric 'parameter space' A. Thus

$$p(.,.,.,.) \in C_b(S \times S \times D \times A)$$

with

$$\sum_j p(i, j, u, \theta) = 1, \quad i \in S, \ u \in D, \ \theta \in A.$$

The actual evolution of the chain will be assumed to correspond to some $\theta_0 \in A$ called the 'true parameter'. The idea is that θ_0 is not known to the controller who tries to estimate it and use the information to devise a suitable control scheme. The self-tuner is one class of such schemes. In our variant thereof, the parameter estimation will be done using the maximum likelihood estimation. We study this next.

The following blanket assumptions will be made:

(A1) S is a single communicating class under all SRS under any θ.

(A2) The empirical process $\{\nu_n\}$ defined in section V.1 is a.s. tight.

Sufficient conditions for (A2) have already been discussed in Chapter V.

(A3) The quantity

$$I\{p(i, j, u, \theta_0) > 0\} \ln(p(i, j, u, \theta)/p(i, j, u, \theta_0))$$

is uniformly bounded in i, j, u, θ and continuous in θ uniformly in i, j, u.

This assumpton is a bit restrictive, and it does appear quite plausible that it can be relaxed considerably at the expense of additional technicalities in the proofs below. We retain it for sake of simplicity.

Let $\{X_n\}$ be a chain governed by a CS $\{\xi_n\}$. The likelihood ratio at time n, $n \geq 1$, is defined by

$$\Lambda_n(\theta) = \prod_{i=0}^{n-1} \frac{p(X_i, X_{i+1}, \xi_i(X_i), \theta)}{p(X_i, X_{i+1}, \xi_i(X_i), \theta_0)}, \quad \theta \in A.$$

The log-likelihood function at time n is given by

$$L_n(\theta) = \ln \Lambda_n(\theta) = \sum_{i=0}^{n-1} \ln(p(X_i, X_{i+1}, \xi_i(X_i), \theta)/p(X_i, X_{i+1}, \xi_i(X_i), \theta_0)), \ \theta \in A.$$

The maximum likelihood estimate (MLE) $\hat{\theta}_n$ of θ_0 at time n is

$$\hat{\theta}_n = \operatorname{argmax} \ \Lambda_n(\theta) = \operatorname{argmax} \ L_n(\theta),$$

where in the case of a nonunique choice of the argmax, one uses some predetermined priority rule to force a unique choice. Let

$$M_n(\theta) = \sum_{i=0}^{n-1} p(X_i, j, \xi_i(X_i), \theta_0) \ln(p(X_i, j, \xi_i(X_i), \theta)/p(X_i, j, \xi_i(X_i), \theta_0))$$

$$= - \sum_{i=0}^{n-1} p(X_i, j, \xi_i(X_i), \theta) \left(\frac{p(X_i, j, \xi_i(X_i), \theta_0)}{p(X_i, j, \xi_i(X_i), \theta)} \ln \left(\frac{p(X_i, j, \xi_i(X_i), \theta_0)}{p(X_i, j, \xi_i(X_i), \theta)} \right) \right). \qquad (1.1)$$

From the convexity of the map $x \to x\ln x$ and Jensen's inequality, it follows that each summand of (1.1) and therefore $M_n(\theta)$ is nonpositive. Letting $\mathcal{F}_n = \sigma(X_i, \xi_i, i \leq n)$, $n \geq 1$, as before, it is easy to see that $L_n(\theta) - M_n(\theta)$, $n \geq 1$, is an $\{\mathcal{F}_n\}$-martingale for each θ. By (A3) above, it has bounded increments.

Lemma 1.1.

$$\lim_{n\to\infty} \frac{L_n(\hat{\theta}_n)}{n} = \lim_{n\to\infty} \frac{M_n(\hat{\theta}_n)}{n} = 0 \text{ a.s.}$$

Proof. Appying the martingale stability theorem of [Lo], p. 53, to $L_n(\theta) - M_n(\theta)$, $n \geq 1$, we have

$$\frac{L_n(\theta) - M_n(\theta)}{n} \to 0 \text{ a.s.}$$

for $\theta \in A$ and hence for θ in a countable dense subset of A. From (A3), it follows that for each sample path, the maps

$$\theta \to \frac{L_n(\theta) - M_n(\theta)}{n}, \quad n \geq 1,$$

are equicontinuous. Thus

$$\sup_{\theta} \left| \frac{L_n(\theta) - M_n(\theta)}{n} \right| \to 0 \text{ a.s.}$$

In particular,

$$\frac{L_n(\hat{\theta}_n) - M_n(\hat{\theta}_n)}{n} \to 0 \text{ a.s.}$$

Since $M_n(\theta) \leq 0$ for $n \geq 1$, $\theta \in A$, we have

$$\limsup_{n \to \infty} \frac{L_n(\hat{\theta}_n)}{n} = \limsup_{n \to \infty} \frac{M_n(\hat{\theta}_n)}{n} \leq 0 \text{ a.s.}$$

On the other hand, the definition of $\hat{\theta}_n$ implies that

$$L_n(\hat{\theta}_n) \geq L_n(\theta_0) = 0, \quad n \geq 1.$$

Thus

$$\liminf_{n \to \infty} \frac{L_n(\hat{\theta}_n)}{n} = \liminf_{n \to \infty} \frac{M_n(\hat{\theta}_n)}{n} \geq 0 \text{ a.s.}$$

The claim follows. QED

Consider the following 'identifiability condition':

(A4) For any $\theta, \theta' \in A$ with $\theta \neq \theta'$ and any $u \in D$, there exist $i, j \in S$ such that

$$p(i, j, u, \theta) \neq p(i, j, u, \theta').$$

Intuitively, (A4) ensures that distinct values of θ will lead to distinct dynamics under any CS and thereby allow the estimation algorithm to discriminate between them. This is made precise below.

Corollary 1.1. Under (A4), $\hat{\theta}_n \to \theta_0$ a.s.

Proof. Let N be the set of zero probability outside which the above lemma holds. Fix any sample path outside N. Let $\bar{\theta}$ be a limit point of $\{\hat{\theta}_n\}$ in A with

$$\hat{\theta}_{n(k)} \to \bar{\theta}$$

for a suitable $\{n(k)\} \subset \{n\}$. (A3) implies that the maps $\theta \to M_n(\theta)/n,\ n \geq 1,$ are equicontinuous. Thus it follows that

$$\lim_{n \to \infty} M_n(\bar{\theta})/n = 0.$$

By (A2) and Lemma V.1.1, this implies that there exists an SSRS $\gamma[\Phi]$, $\Phi = \prod_i \hat{\Phi}_i$, such that

$$\sum_i \pi[\Phi](i) \int \left(\sum_j p(i, j, u, \theta) \left(\frac{p(i, j, u, \theta_0)}{p(i, j, u, \theta)} \ln \left(\frac{p(i, j, u, \theta_0)}{p(i, j, u, \theta)} \right) \right) \right) \hat{\Phi}_i(du) = 0$$

Using Jensen's inequality and the strict convexity of the map $x \to x\ln x$, a standard argument shows that one must have

$$p(i, j, u, \bar{\theta}) = p(i, j, u, \theta_0)$$

for all $i, j \in S$ and $\hat{\Phi}_i$ - a.s. u. This contradicts (A4) unless $\bar{\theta} = \theta_0$, proving the claim.

QED

In the language of statistics, we have proved that the MLE are consistent in the strong sense.

IX.2 Parameter Dependence of the Value Function

Recall the value function $V = V\{\xi_0\}$ for the ergodic control problem from Chapter VI, $\gamma\{\xi_0\}$ being any optimal SSS. Now that our transition matrices depend on the additional parameter θ, so will V. We shall make this θ-dependence explicit by writing $V = V^{\theta}$. (We shall similarly denote the θ-dependence elsewhere by means of a superscript without explicitly mentioning this notational change on every occasion.) The dynamic programming equations now become

$$\beta^{\theta} 1_c = \min_{\xi} ((P^{\theta}\{\xi\} - I)V^{\theta} + k\{\xi\}) \tag{2.1a}$$

$$= \min_{\Phi} ((P^{\theta}[\Phi] - I)V^{\theta} + k[\Phi]). \tag{2.1b}$$

Throughout this chapter, we assume that k is bounded. We prove below the continuity of the map $\theta \to V^{\theta}$ at $\theta = \theta_0$ under three sets of hypotheses. Each of these hypotheses allows us to derive (2.1 a-b) and to show that an SSS $\gamma\{\xi_{\theta}\}$ is optimal under θ if and only if ξ_{θ} attains the minimum in (2.1a). This, along with the selection theorem of Lemma 1, [Ben], allows us to pick a $\xi_{\theta} \in L$ for each $\theta \in A$ such that $\gamma\{\xi_{\theta}\}$ is an optimal SSS under θ and the map $\theta \to \xi_{\theta} : A \to L$ is measurable.

The three cases we consider are: (1) the Liapunov condition, (2) the strong uniform recurrence hypothesis and (3) the near-monotonicity. We consider these one by one.

Case 1: The Liapunov Condition

Assume the following:

(A5) There exists a function $w : S \to R^+$ such that

(i) $\lim_{i \to \infty} w(i) = \infty$,

(ii) there exist $a > 0$, $\varepsilon_0 > 0$ such that

$$E^{\theta}[(w(X_{n+1}) - w(X_n) + \varepsilon_0) I\{w(X_n) > a\} / \mathcal{F}_n] \leq 0, \ n \geq 1,$$

under any CS and any θ, and

(iii) there exists a random variable Z and a scalar $\lambda > 0$ such that

$$E[\exp(\lambda Z)] < \infty$$

and for all $c \in R$,

$$P^{\theta}(|w(X_{n+1}) - w(X_n)| > c) \leq P(Z > c), \ n \geq 0,$$

under any CS and any $\theta \in A$.

This is the Liapunov condition of section V.3, assumed to hold uniformly with respect to θ. We also assume the following:

(A6) For each $i \in S$, there exists a finite set $R_i \subset S$ such that $p(i, j, u, \theta) = 0$ for all u, θ whenever $j \notin R_i$.

We start with some preliminaries. Let $p^n_\Phi(i, j, \theta)$ denote the (i,j)-th element of $(P^\theta[\Phi])^n$ for $n \geq 1$, $i, j \in S$. Fix $i, j \in S$ and define

$$n^\theta[\Phi] = \min \{n \geq 1 \mid p^n_\Phi(i, j, \theta) > 0\}.$$

Lemma 2.1. There exists a finite integer $N \geq 1$ such that for any $\theta \in A$ and any SRS $\gamma[\Phi]$, $n^\theta[\Phi] \leq N < \infty$.

Proof. Suppose not. Then for every integer $N \geq 1$, there exists an SRS $\gamma[\Phi(N)]$ and a $\theta(N) \in A$ such that

$$p^m_{\Phi(N)}(i, j, \theta(N)) = 0, \quad 1 \leq m \leq N.$$

Let $(\Phi(\infty), \theta(\infty))$ be a limit point of $(\Phi(N), \theta(N))$ in $\mathbb{P}_0(L) \times A$ as $N \to \infty$. By the continuity of $p(.,.,.,.)$ and (A6) above, it follows that the map $(\Phi, \theta) \in \mathbb{P}_0(L) \times A \to p^n_\Phi(i,j,\theta) \in [0,1]$ is continuous for each $n \geq 1$. Thus we must have

$$p^n_{\Phi(\infty)}(i, j, \theta(\infty)) = 0, \quad n \geq 1,$$

contradicting (A1). This proves the claim. QED

Corollary 2.1. There exists a $\rho > 0$ such that

$$\sup_{1 \leq n \leq N} p^n_\Phi(i, j, \theta) \geq \rho > 0$$

for all SRS $\gamma[\Phi]$ and all $\theta \in A$, N being as in the preceeding lemma.

Proof. Suppose not. Then there exists a sequence $(\Phi(n), \theta(n))$ in $\mathbb{P}_0(L) \times A$ such that

$$p^m_{\Phi(n)}(i, j, \theta(n)) \to 0, \quad 1 \leq m \leq N,$$

as $n \to \infty$. Since the map

$$(\Phi, \theta) \to [p^1_\Phi(i, j, \theta), p^2_\Phi(i, j, \theta), \ldots, p^N_\Phi(i, j, \theta)]$$

is continuous as observed above and its domain $\mathbb{P}_0(L) \times A$ is compact, so must be its range. Hence $[0,0,\ldots,0]$ is in its range, contradicting Lemma 2.1. The claim follows. QED

Lemma 2.2. For each $i \in S$,

$$\sup_{\theta} | V^{\theta}(i) | < \infty.$$

Proof. Fix $i, \ell \in S$. From (2.1), we have

$$V^{\theta}(i) + \beta^{\theta} = \min_{u} [\, k(i, u) + \sum_{j} p(i, j, u, \theta) V^{\theta}(j) \,]$$

$$= k(i, \xi_{\theta}(i)) + \sum_{j} p(i, j, \xi_{\theta}(i), \theta) V^{\theta}(j)$$

$$\geq \sum_{j} p(i, j, \xi_{\theta}(i), \theta) V^{\theta}(j).$$

Iterating and using the above lemma,

$$V^{\theta}(i) \geq \rho V^{\theta}(\ell) - \sup_{\theta} \beta^{\theta}.$$

for some $\rho > 0$. If we can find $\{\theta(n)\} \subset A$ such that $V^{\theta(n)}(\ell) \to \infty$, then $V^{\theta(n)}(i) \to \infty$ as well. Similarly, if there exist $\{\theta(n)\} \subset A$ such that $V^{\theta(n)}(i) \to -\infty$, then $V^{\theta(n)}(\ell) \to -\infty$. Given the arbitrary choice of i, ℓ, it follows that $\left\{ | V^{\theta(n)}(i) |, \theta \in A \right\}$ is unbounded for some $i \in S$ if and only if it is so far all i. Since $V^{\theta}(1) = 0$ for all $\theta \in A$, the claim follows. QED

Corollary 2.2. There exists $C_1, C_2 > 0$ such that

$$| V^{\theta}(i) | \leq C_1 + C_2\, w(i), \quad i \in S, \ \theta \in A.$$

Proof. This follows as in Lemma VI.4.3–VI.4.4 in view of the above lemma. QED

Theorem 2.1. The map $\theta \to V^{\theta} : A \to R^{\infty}$ is continuous.

Proof. Let $\theta(n) \to \theta(\infty)$ in A. In view of Lemma 2.2, we may drop to a subsequence if necessary and suppose that

$$V^{\theta(n)} \to \bar{V} \ \text{in} \ R^{\infty},$$

$$\beta^{\theta(n)} \to \bar{\beta} \ \text{in} \ R.$$

for some $\bar{V} \in R^{\infty}$, $\bar{\beta} \in R$. (A6) allows us to pass to the limit in (2.1a) for $\theta = \theta(n)$, to obtain the same equation with $(\bar{V}, \bar{\beta}, P^{\theta(\infty)}\{\xi\})$ replacing $(V^{\theta}, \beta^{\theta}, P^{\theta}\{\xi\})$. Passing to the limit in

$$| V^{\theta(n)}(i) | \leq C_1 + C_2 w(i), \ i \in S, \ n \geq 1,$$

we have

$$| \bar{V}(i) | \leq C_1 + C_2 w(i), \ i \in S.$$

Clearly, $\bar{V}(1) = 0$. From Theorem VI.4.2, we then have $\bar{V} = V^{\theta(\infty)}$ and $\bar{\beta} = \beta^{\theta(\infty)}$, proving the claim. QED

Case 2 :The strong Uniform Recurrence Hypothesis

Instead of (A5) and (A6), assume the following:

(A7)
$$\sup_{\theta} \sup_{i} \sup_{\gamma\{\xi\}} E_{\xi}^{\theta}[\tau(1)/X_0 = i] < \infty. \tag{2.2}$$

This is the uniform recurrence condition (V.3.8) of Chapter V, assumed to hold uniformly in θ. Since

$$V^{\theta}(i) = \min_{\gamma\{\xi\}} E^{\theta}\Big[\sum_{n=0}^{\tau(1)-1} (k(X_n, \xi(X_n)) - \beta^{\theta})/X_0 = i \Big], \ i \in S,$$

and k is bounded, we have

$$K \overset{\Delta}{=} \sup_{i} \sup_{\theta} | V^{\theta}(i) | < \infty \tag{2.3}$$

Theorem 2.2. The map $\theta \rightarrow V^{\theta} : A \rightarrow R^{\infty}$ is continuous.

Proof. Let $\theta(n) \rightarrow \theta(\infty)$ in A. In view of (2.3), we may drop to a subsequence if necessary and suppose that $(V^{\theta(n)}, \beta^{\theta(n)}, \xi_{\theta(n)})$ converge to some $(\bar{V}, \bar{\beta}, \bar{\xi})$ in $R^{\infty} \times R \times L$. Since

$$p(i, j, \xi_{\theta(n)}(i), \theta(n)) \rightarrow p(i, j, \bar{\xi}(i), \theta(\infty))$$

for each i,j, Scheffe's theorem implies that

$$p(i,., \xi_{\theta(n)}(i), \theta(n)) \rightarrow p(i,., \bar{\xi}(i), \theta(\infty)) \tag{2.4}$$

in total variation for each $i \in S$. Now for each $i \in S$.

$$
\begin{aligned}
&|\min_u \Big(\sum_j p(i, j, u, \theta(n))\, V^{\theta(n)}(j) + k(i, u)\Big) \\
&\quad - \big(\Sigma\, p(i, j, \bar{\xi}(i), \theta(\infty))\bar{V}(j) + k(i, \bar{\xi}(i))\big)| \\
&\quad \le |\sum_j \big(p(i, j, \xi_{\theta(n)}(i), \theta(n)) - p(i, j, \bar{\xi}(i), \theta(\infty))\big)\, V^{\theta(n)}(j)| \\
&\quad + |\sum_j p(i, j, \bar{\xi}(i), \theta(\infty))(V^{\theta(n)}(j) - \bar{V}(j))| \\
&\quad + |k(i, \xi_{\theta(n)}(i)) - k(i, \bar{\xi}(i))| \\
&\quad \le K \,\| p(i,., \xi_{\theta(n)}(i), \theta(n)) - p(i,., \bar{\xi}(i), \theta(\infty)) \|_{TV} \\
&\quad + \sum_j p(i, j, \bar{\xi}(i), \theta(\infty))\, |V^{\theta(n)}(j) - \bar{V}(j)| \\
&\quad + |k(i, \xi_{\theta(n)}(i)) - k(i, \bar{\xi}(i))|
\end{aligned}
$$

where $\|\cdot\|_{TV}$ is the total variation norm. As $n \to \infty$, the first term on the right goes to zero by (2.4), the second by (2.3) and the dominated convergence theorem, and the third by virtue of the continuity of k. A similar argument also shows that for any $\xi \in L$,

$$
|\Big(\sum_j p(i, j, \xi(i), \theta(n))V^{\theta(n)}(j) + k(i, \xi(i))\Big) - \Big(\sum_j p(i, j, \xi(i), \theta(\infty))\bar{V}(j) + k(i, \xi(i))\Big)| \to 0
$$

as $n \to \infty$. Since

$$
\sum_j p(i, j, \xi(i), \theta(n))V^{\theta(n)}(j) + k(i, \xi(i)) \ge \min_u \Big(\sum_j p(i, j, u, \theta(n))V^{\theta(n)}(j) + k(i, u)\Big),
$$

we can let $n \to \infty$ to obtain

$$
\sum_j p(i, j, \xi(i), \theta(\infty))\bar{V}(j) + k(i, \xi(i)) \ge \sum_j p(i, j, \bar{\xi}(i), \theta(\infty))\bar{V}(j) + k(i, \bar{\xi}(i)).
$$

Since $\xi \in L$ was arbitrary, this shows that

$$\sum_j p(i, j, \bar{\xi}(i), \theta(\infty))\bar{V}(j) + k(i, \bar{\xi}(i)) = \min_u \Big(\sum_j p(i, j, u, \theta(\infty))\, \bar{V}(j) + k(i, u)\Big).$$

We have proved:

$$\Big| \min_u \Big(\sum_j p(i, j, u, \theta(n))V^{\theta(n)}(j) + k(i, u)\Big) - \min_u \Big(\sum_j p(i, j, u, \theta(\infty))\bar{V}(j) + k(i, u)\Big) \Big| \to 0$$

as $n \to \infty$. Thus we may pass to the limit as $n \to \infty$ in (2.1a) for $\theta = \theta(n)$, to obtain the same equation with $(\bar{V}, \bar{\beta}, P^{\theta(\infty)}\{\xi\})$ replacing $(V^{\theta}, \beta^{\theta}, P^{\theta}\{\xi\})$. From (2.3), one has $\sup_i |\bar{V}(i)| < \infty$. Clearly, $\bar{V}(0) = 0$. From Theorem VI.4.3 one then has $\bar{V} = V^{\theta(\infty)}$, $\bar{\beta} = \beta^{\theta(\infty)}$. The claim follows. QED

Case 3: The near-Monotonicity Condition

The third set of conditions we consider is the following:

(A8) $$\liminf_{i \to \infty} \inf_u k(i, u) > \sup_\theta \beta^{\theta}.$$

This is simply the near-monotonicity condition of Chapter V, assumed to hold uniformly in θ. In addition, we shall assume (A6) and the following.

(A9) There exists a function $\bar{w} : S \to R^+$ and an open neighbourhood B of θ_0 such that

(i) $\lim_{i \to \infty} \bar{w}(i) = \infty$.

(ii) There exists $\bar{a}, \bar{\varepsilon} > 0$ such that for each $\theta \in B$,

$$E^{\theta}[(\bar{w}(X_{n+1}) - \bar{w}(X_n) + \bar{\varepsilon})\, I\{\bar{w}(X_n) > \bar{a}\} / \mathcal{F}_n] \le 0, \ \ n \ge 0,$$

under $\gamma\{\xi_{\theta_0}\}$.

(iii) There exists a random variable $\bar{Z}$ and a scalar $\bar{\lambda} > 0$ such that

$$E[\exp(\bar{\lambda}\bar{Z})] < \infty$$

and for any $c \in R^+$, $\theta \in B$,

$$P^{\theta}(|\bar{w}(X_{n+1}) - \bar{w}(X_n)| > c) \leq P(\bar{Z} \geq c), \ n \geq 0,$$

under $\gamma\{\xi_{\theta_0}\}$.

This is again a Liapunov-type condition except that now it is required to hold only under the optimal SSS $\gamma\{\xi_{\theta_0}\}$ and for $\theta \in B$.

Theorem 2.3. The map $\theta \to V^{\theta} : A \to R^{\infty}$ is continuous at θ_0.

Proof. Let $\theta(n) \to \theta_0$ in A. Lemma 2.2 continues to hold here. Thus we may drop to a subsequence if necessary and assume that $(V^{\theta(n)}, \beta^{\theta(n)})$ converges to some $(\bar{V}, \bar{\beta})$ in $R^{\infty} \times R$ as $n \to \infty$. Argue as in Theorem 2.1 to conclude that $(\bar{V}, \bar{\beta})$ satisfy (2.1a). In view of (A8), the set

$$G = \{i \in S \mid \inf_u k(i, u) \leq \sup_{\theta} \beta^{\theta} + \varepsilon\}$$

is finite non-empty for some $\varepsilon > 0$. Then for $\tau = \min\{n \geq 1 \mid X_n \in G\}$,

$$V^{\theta(n)}(i) = E^{\theta(n)}\Big[\sum_{m=0}^{\tau-1} (k(X_m, \xi_{\theta(n)}(X_m)) - \beta) + V^{\theta(n)}(X_{\tau})/X_0 = i\Big],$$

$$\geq \min_{j \in G} V^{\theta(n)}(j)$$

for $n \geq 1$. Leting $n \to \infty$,

$$\bar{V}(i) \geq \min_{j \in G} \bar{V}(j), \ i \in S,$$

implying that $\bar{V}$ is bounded from below. Clearly $\bar{V}(0) = 0$. One may argue as for the Liapunov condition of section V.3 using (A9) in its place, to conclude that $\pi^{\theta(n)}\{\xi_{\theta_0}\}$, $n \geq 1$, are tight. Let π denote any limit point thereof. Letting $n \to \infty$ along an appropriate subsequence in the equation

$$\pi^{\theta(n)}\{\xi_{\theta_0}\}P^{\theta(n)}\{\xi_{\theta_0}\} = \pi^{\theta(n)}\{\xi_{\theta_0}\}$$

and arguing as in the proof of Corollary V.1.1, one has

$$\pi P^{\theta_0}\{\xi_{\theta_0}\} = \pi,$$

that is, $\pi = \pi^{\theta_0}\{\xi_{\theta_0}\}$. Thus $\pi^{\theta(n)}\{\xi_{\theta_0}\} \to \pi^{\theta_0}\{\xi_{\theta_0}\}$ and

$$\bar{\beta} \leftarrow \beta^{\theta(n)} \le \sum_i k(i, \xi_{\theta_0}(i))\pi^{\theta(n)}\{\xi_{\theta_0}\}(i)$$

$$\to \sum_i k(i, \xi_{\theta_0}(i))\pi^{\theta_0}\{\xi_{\theta_0}\}(i) = \beta^{\theta_0}.$$

Thus $\bar{\beta} \le \beta^{\theta_0}$. By Theorem VI.4.1, $\bar{V} = V^{\theta_0}$, $\bar{\beta} = \beta^{\theta_0}$ and the proof is completed. QED

IX.3 Optimality of the Self-Tuner

In the self-tuning scheme, the controller uses at each time a control that would have been optimal if the current parameter estimate were the true parameter. In our set-up, this amounts to using the CS $\{\xi_n\}$ given by

$$\xi_n = \xi_{\hat{\theta}_n}, \quad n \ge 1. \tag{3.1}$$

We prove below the a.s. optimality of this CS for the three cases considered above under the assumptions (A1), (A3), the 'identifiability condition' (A4) and in case of Case 3 above, under the following additional assumption.

(A10) There exists a function $w' : S \to R^+$ such that:

(i) $\lim_{i \to \infty} w'(i) = \infty$.

(ii) There exist $a' > 0$, $\varepsilon' > 0$ such that

$$E^{\theta_0}[\,(w'(X_{n+1}) - w'(X_n) + \varepsilon')\, I\{w'(X_n) > a'\} / \mathcal{F}_n] \le 0, \quad n \ge 0,$$

under any $\gamma\{\xi_\theta\}$, $\theta \in A$.

(iii) There exist a random variable Z' and a scalar $\lambda' > 0$ such that

$$E[\,\exp(\lambda' Z')\,] < \infty$$

and for all $c \in R^+$,

$$P^{\theta_0}(|W'(X_{n+1}) - W'(X_n)| > c) \le P(Z' > c), \ n \ge 0,$$

under any $\gamma\{\xi_\theta\}$, $\theta \in A$.

Lemma 3.1. (A2) holds.

Proof. For Case 1 and Case 2 above, we already know this from Chapter V. As for Case 3, it is reduced to Case 1 by (3.1) coupled with (A10) above. QED

Corollary 3.1. $\hat{\theta}_n \to \theta_0$ a.s.

For $i \in S$, $u \in D$, $\theta \in A$, let

$$F(i, u, \theta) = k(i, u) + \sum_j p(i, j, u, \theta)V^\theta(j).$$

Let

$$U_{i,\theta} = \{u \in D \mid F(i, u, \theta) = \min_{v \in D} F(i, v, \theta)\}. \tag{3.2}$$

By our characterization of the optimal SSS in Chapter VI, we have $\xi_\theta(i) \in U_{i,\theta}$ for each i, θ, for all three cases above. For $\varepsilon > 0$, let $U^\varepsilon_{i,\theta}$ be the closed ε-neighbourhood of $U_{i,\theta}$.

Lemma 3.2. For each $i \in S$, $\xi_n(i) \to U_{i,\theta_0}$ a.s.

Proof. For $i \in S$, $n \ge 1$, $\xi_n(i) \in U_{i,\hat{\theta}_n}$ by (3.1) and the remark following (3.2). Thus

$$F(i, \xi_n(i), \hat{\theta}_n) \le F(i, u, \hat{\theta}_n), \ u \in D.$$

In view of Corollary 3.1 and Theorems 2.1 - 2.3, it follows that for almost all sample paths, any limit point ξ of $\{\xi_n\}$ in L must satisfy

$$F(i, \xi(i), \theta_0) \le F(i, u, \theta_0)$$

for u in a countable dense set of D, and hence for all $u \in D$ by continuity. The claim follows. QED

Theorem 3.1. The CS $\{\xi_n\}$ of (3.1) is optimal.

Proof. Let N be the set of zero probability outside which the conclusions of Lemma 3.2 (for all $i \in S$) and Lemma V.1.1 hold. Fix a sample path outside N. Let ν be a limit point of $\{\nu_n\}$ in $P(\bar{S} \times D)$. Recall (A2). By (A2), $\delta_\nu = 0$ in Lemma V.1.1. Thus ν is of the form

$$\nu(\{i\}, du) = \pi^{\theta_0}[\Phi](i)\hat{\Phi}_i(du), \quad i \in S,$$

for some SSRS $\gamma[\Phi]$, $\Phi = \prod_i \hat{\Phi}_i$. Let $\varepsilon > 0$. By Lemma 3.2, for each $i \in S$, $\xi_{\hat{\theta}_n}(i) \in U^{\varepsilon}_{i,\theta_0}$ for n sufficiently large. For $\ell \geq 1$,

$$\nu_n\Big(\bigcup_{i \in S} \{i\} \times U^{\varepsilon}_{i,\theta_0}\Big) = \frac{1}{n} \sum_{m=0}^{n-1} \sum_{i \in S} I\{X_m = i, \xi_{\hat{\theta}_m}(i) \in U^{\varepsilon}_{i,\theta_0}\}$$

$$\geq \frac{1}{n} \sum_{m=0}^{n-1} \sum_{i=0}^{\ell} I\{X_m = i, \xi_{\hat{\theta}_m}(i) \in U^{\varepsilon}_{i,\theta_0}\}.$$

Let $n \to \infty$ along a subsequence along which $\nu_n \to \nu$. Then

$$\nu\Big(\bigcup_{i \in S} \{i\} \times U^{\varepsilon}_{i,\theta_0}\Big) \geq \lim_{n \to \infty} \frac{1}{n} \sum_{m=0}^{n-1} \sum_{i=0}^{\ell} I\{X_m = i, \xi_{\theta_m}(i) \in U^{\varepsilon}_{i,\theta_0}\}$$

$$= \sum_{i=0}^{\ell} \lim_{n \to \infty} \frac{1}{n} \sum_{m=0}^{n-1} I\{X_m = i, \xi_{\hat{\theta}_m}(i) \in U^{\varepsilon}_{i,\theta_0}\}$$

$$= \sum_{i=0}^{\ell} \lim_{n \to \infty} \frac{1}{n} \sum_{m=0}^{n-1} I\{X_m = i\},$$

since $\xi_{\hat{\theta}_n}(i) \in U^{\varepsilon}_{i,\theta_0}$ eventually for each i. Thus

$$\nu\Big(\bigcup_{i \in S} \{i\} \times U^{\varepsilon}_{i,\theta_0}\Big) \geq \nu(\{0,1,\ldots,\ell\} \times D).$$

Letting $\ell \to \infty$, we get

$$\nu\Big(\bigcup_{i \in S} \{i\} \times U^{\varepsilon}_{i,\theta_0}\Big) \geq 1.$$

Let $\varepsilon \downarrow 0$. Then

$$\nu(\bigcup_{i \in S} \{i\} \times U_{i,\theta_0}) = 1.$$

Thus $\hat{\Phi}_i$ is concentrated on U_{i,θ_0} for each i. Argue as in Theorem VI.3.1 to conclude that $\gamma[\Phi]$ is an optimal SRS, that is,

$$\int k d\nu = \beta^{\theta_0}.$$

Since ν was any limit point of $\{\nu_n\}$, the claim follows. QED

IX.4. Dropping the Idenifiability Condition

The validity of Theorem 3.1 above depends crucially on the identifiability condition (A4). In this section we see some consequences of dropping this condition. We shall consider finite S, A and D and show that $\{\hat{\theta}_n\}$ converges a.s., albeit not necessarily to θ_0. The characterization of possible limit points of $\{\hat{\theta}_n\}$ gives a clue as to what can go wrong.

Let $A = \{\theta_0, \theta_1, \ldots, \theta_M\}$. Then for each i, $0 \leq i \leq M$, $(\Lambda_n(\theta_i), \mathcal{F}_n)$ is a non-negative martingale and hence converges a.s. Let N be the set of zero probability outside which this convergence holds for all i, $0 \leq i \leq M$. Fix a sample point outside N and let $\bar{\theta}$ be a limit point of $\{\hat{\theta}_n\}$. Since A is finite, $\hat{\theta}_n = \bar{\theta}$ infinitely often. Thus $\Lambda_n(\bar{\theta}) \geq \Lambda_n(\theta_0) = 1$ infinitely often, implying

$$\Lambda_\infty(\bar{\theta}) \stackrel{\Delta}{=} \lim_{n \to \infty} \Lambda_n(\bar{\theta}) \geq 1.$$

Thus

$$\frac{p(X_n, X_{n+1}, \xi_n(X_n), \bar{\theta})}{p(X_n, X_{n+1}, \xi_n(X_n), \theta_0)} = \frac{\Lambda_{n+1}(\bar{\theta})}{\Lambda_n(\bar{\theta})} \to 1. \quad (5.1)$$

Since the left hand side is finite-valued, it actually equals one for n sufficiently large. Thus $\Lambda_n(\bar{\theta}) = \Lambda_\infty(\bar{\theta})$ for sufficiently large n. Since this is so for any limit point $\bar{\theta}$ of $\{\hat{\theta}_n\}$ and there are at most finitely many of them, it must hold for all of them for sufficiently large n, say $n \geq n_0$. By increasing n_0 if necessary, assume that $\hat{\theta}_n$ is already in the set A' of limit points of $\{\hat{\theta}_m\}$ for $n \geq n_0$. (Again, this is possible because A is finite.) Then for $n \geq n_0$, the MLE would be the maximizer of $\theta \in A' \to \Lambda_\infty(\theta)$, resolving any tie by a preassigned priority rule. It is clear that the choice of $\hat{\theta}_n$ must remain the same for all $n \geq n_0$. In other words, $\hat{\theta}_n \to \theta^*$ for some $\theta^* \in A$ and $\hat{\theta}_n = \theta^*$ for sufficiently large n. We characterize this θ^* below.

Theorem 4.1. $\hat{\theta}_n \to \theta^*$ a.s. for an A-valued random variable θ^* satisfying:

$$p(i, j, \xi_{\theta^*}(i), \theta^*) = p(i, j, \xi_{\theta^*}(i), \theta_0), \quad i, j \in S. \tag{4.2}$$

Proof. The first part is already proved. From Lemma 1.1, we have

$$\lim_{n\to\infty} \frac{1}{n} \sum_{m=0}^{n-1} \Big(\sum_{j\in S} p(X_m, j, \xi_{\hat{\theta}_m}(X_m), \hat{\theta}_m)$$

$$\ln(p(X_m, j, \xi_{\hat{\theta}_m}(X_m), \hat{\theta}_m)/p(X_m, j, \xi_{\hat{\theta}_m}(X_m), \theta_0)))$$

$$= \lim_{n\to\infty} \frac{1}{n} \sum_{m=0}^{n-1} \Big(\sum_{j\in S} p(X_m, j, \xi_{\theta^*}(X_m), \theta^*)$$

$$\ln(p(X_m, j, \xi_{\theta^*}(X_m), \theta^*)/p(X_m, j, \xi_{\theta^*}(X_m), \theta_0))) = 0 \text{ a.s.}, \tag{4.3}$$

where we have exploited the fact that $\hat{\theta}_n = \theta^*$ from some n onwards, a.s. Let N be the set of zero probability above, enlarged if necessary so that (4.3) and Lemma V.1.1 hold outside N. Fix a sample path outside N. From Lemma V.1.1 and the fact that $\xi_{\hat{\theta}_n} = \xi_{\theta^*}$ from some n onwards, it follows that

$$\nu_n \to \hat{\pi}^{\theta_0}\{\xi_{\theta^*}\}. \tag{4.4}$$

(Note that $\delta_\nu = 0$ in Lemma V.1.1 because S is finite.) (4.3) and (4.4) together imply

$$\sum_i \pi^{\theta_0}\{\xi_{\theta^*}\}(i) \sum_j p(i, j, \xi_{\theta^*}(i), \theta^*) \left(\frac{p(i,j,\xi_{\theta^*}(i),\theta_0)}{p(i,j,\xi_{\theta^*}(i),\theta^*)} \ln \left(\frac{p(i,j,\xi_{\theta^*}(i),\theta_0)}{p(i,j,\xi_{\theta^*}(i),\theta^*)} \right) \right) = 0.$$

From the Jensen's inequality and strict convexity of the map $x \to x \ln x$, a standard argument yields (4.2). QED

Corollary 4.1. Under (3.1),

$$\frac{1}{n}\sum_{m=0}^{n-1} k(X_m, \xi_m(X_m)) \to \beta^{\theta^*} \text{ a.s.}$$

for θ^* as in Theorem 4.1.

Proof. From (4.4), we have

$$\frac{1}{n}\sum_{m=0}^{n-1} k(X_m, \xi_m(X_m)) \to \sum_i \pi^{\theta_0}\{\xi_{\theta^*}\}(i)k(i, \xi_{\theta^*}(i)).$$

From (4.2), it follows that $\pi^{\theta_0}\{\xi_{\theta^*}\} = \pi^{\theta^*}\{\xi_{\theta^*}\}$. Thus the above limit coincides with β^{θ^*}.

QED

This does not, however, imply that $\beta^{\theta^*} = \beta^{\theta_0}$. Thus the CS (3.1) need not be optimal. See [KuBe] for a concrete example of a situation where (A4) fails and (3.1) is not optimal.

Nevertheless, all is not lost. There are various modifications of the 'raw self-tuner' discussed here that allow one to circumvent the absence of (A4). These are: randomization of the control or the parameter estimate [BoV3], and introduction of the cost bias in the estimation scheme [KuBe], [KuLi], [Bor 9]. The latter will be discussed at length in the next chapter.

Chapter X
Adaptive Control: The Kumar-Becker-Lin Scheme

Among the various modifications of the raw self-tuner described in the preceding chapter, the Kumar-Becker-Lin scheme of [KuBe], [KuLi] is perhaps the most appealing. It circumvents the need to assume the identifiability condition by introducing a cost-bias in the log-likelihood function. This favours parameters leading to a lower optimal cost. We describe this scheme in the first section below, along with a technical aside for later use. The second section studies the asymptotic behaviour of the so-called 'cost-biased maximum likelihood estimates' (CMLE). The last section uses these results to prove the a.s. optimality of the KBL scheme. The main reference is [Bor 9].

X.1 Preliminaries

We assume from now on that the running cost $k(.,.)$ is bounded and bounded away from zero from below. The latter assumption is not restrictive because if it were not, one could always make it so by adding a suitable constant. It implies in particular that β^{θ}, $\theta \in A$, is bounded away from zero from below, allowing us to define the cost-biased log-likelihood function $\tilde{L}_n(\theta)$, $n \geq 1$, $\theta \in A$, by

$$\tilde{L}_n(\theta) = L_n(\theta) - n^{\alpha} \ln(\beta^{\theta}/\beta^{\theta_0}), \quad n \geq 1, \quad \theta \in A,$$

for some prescribed $\alpha \in (1/2, 1)$. The cost-biased maximum likelihood estimate (CMLE) at time n will be defined as

$$\tilde{\theta}_n = \operatorname{argmax}(\tilde{L}_n(\theta)), \quad n \geq 1,$$

where a tie is resolved according to some predetermined priority rule so as to force a unique choice of $\tilde{\theta}_n$. Let $\{y_i\}$ be a prescribed increasing sequence of positive integers such that

$$\sum_n y_n^{-\ell} < \infty \tag{1.1}$$

for some $\ell \geq 1$. Given a controlled Markov chain $\{X_n\}$ as in the last chapter, we define the stopping times τ_n, $n \geq 1$, by $\tau_1 = 0$ and

$$\tau_n = \min\{m > \tau_{n-1} \mid X_m = 1 \text{ or } m = \tau_{n-1} + y_n\},\ n \geq 2.$$

For $n \geq 1$, let $\sigma(n)$ = the largest τ_i not exceeding n. The variant of the KBL scheme which we study here is given by the CS $\{\xi_n\}$ where

$$\xi_n = \xi_{\hat{\theta}_{\sigma(n)}},\ n \geq 1. \tag{1.2}$$

As already mentioned, the cost bias favours parameters leading to lower optimal costs, thereby aiding the parameter selection. Our replacement of n by $\sigma(n)$ indicates that any candidate for the unknown parameter is tried out for a time sufficient for the algorithm to test it. We prove the a.s. optimality of this scheme in the next two sections. The rest of this section is devoted to a digression into a classification of the limit points of a sequence taking values in a compact metric space. This will be useful later.

Let X be a compact metric space and $\{x_n\}$ a sequence therein. Let A be the set of limit points of $\{x_n\}$. Then A is compact non-empty. Call $x \in A$ a frequent limit point of $\{x_n\}$ if for any open neighbourhood B of x in X,

$$\limsup_{n\to\infty} \frac{1}{n} \sum_{m=1}^{n} I\{x_m \in B\} > 0.$$

An $x \in A$ which is not a frequent limit point of $\{x_n\}$ will be said to be its rare limit point. Similarly, one characterizes the events $\{A_n\}$ as frequent or rare according to whether

$$\limsup_{n\to\infty} \frac{1}{n} \sum_{m=1}^{n} I_{A_m} > 0 \text{ or } = 0.$$

Lemma 1.1. The set A^* of frequent limit points of $\{x_n\}$ is compact nonempty , and for any open set $G \supset A^*$, the events $\{x_n \notin G\}$ are rare.

Proof. Let $\varepsilon > 0$ and cover A with finitely many balls of radius ε, say $B_1,\ldots,B_{m_1}$. Then $\{x_n\}$ is eventually in $\cup B_i$ and for at least one i, say $i = k$, the events $\{x_n \in B_k\}$ are frequent. Cover $\bar{B}_k$ by finitely many balls of radius $\varepsilon/2$, say $B_{k1},\ldots,B_{km_2}$. Then for some j, the events $\{x_n \in B_k \cap B_{kj}\}$ are frequent. Cover $\overline{B_k \cap B_{kj}}$ by finitely many balls of radius $\varepsilon/4$. Continuing in this manner, we form a sequence of balls of radius $\varepsilon/2^n$, $n = 1,2,\ldots$, such that the events e_{x_n} belongs to any prescribed one of them are frequent. The centres of these balls will have a limit point x which can easily be shown to be a frequent limit point. Thus A^* is nonempty. If $\bar{x} \in \overline{A^*}$ and B is an open neighbourhood of $\bar{x}$, then there exist $x' \in A^*$ and an open neighoburhood B' of x' such that $x' \in B' \subset B$. Since the events

$\{x_n \in B'\}$ are frequent, so are the events $\{x_n \in B\}$. Since B was an arbitrary open neighbourhood of $\bar{x}$, this proves that $\bar{x}$ is a frequent limit point. Thus A^* is closed and hence compact. Let $G' = G^c \cap A$. Then G' is compact and $G' \cap A^* = \emptyset$. Let G_1, G_2 be disjont open sets such that $G' \subset G_1$, $A^* \subset G_2$. For $\varepsilon > 0$, let $D'_1,\ldots,D'_m$ be balls of radius ε which cover G'. Then $D_1 = D'_1 \cap G_1,\ldots,D_m = D'_m \cap G_1$ is a finite cover of G' which does not intersect A^*. If $\{x_n \in G^c\}$ were frequent, so would be $\{x_n \in D_k\}$ for some k. Proceed as in the proof of the first part of this lemma to conclude that we can find a frequent limit point of $\{x_n\}$ in G'. This contradicts $G' \cap A^* = \emptyset$. Hence the events $\{x_n \notin G\}$ are rare. QED

Lemma 1.2. Let $\{n_k, k \geq 1\}$ be an increasing sequence of positive integers such that

$$\limsup_{n \to \infty} \frac{1}{n} \sum_{m=1}^{n} I\{m = n_k \text{ for some } k\} > 0. \tag{1.3}$$

Let $\{x_n\} \subset X$ be as above. Then the subsequence $\{x_{n_k}, k \geq 1\}$ has a limit point which is a frequent limit point of $\{x_n\}$.

Proof. Let A' denote the set of limit points of $\{x_{n_k}\}$. For $\varepsilon > 0$, cover A' by finitely many balls of radius ε, say $B_1,\ldots,B_m$. Then $\{x_{n_k}\}$ is in $\cup B_i$ from some k onwards. By (1.3), it follows that the events $\{x_n \in \cup B_i\}$ are frequent. Proceed as in the proof of the above lemma to conclude. QED

X.2 Asymptotic Behaviour of the CMLE

From now on, we shall assume that A is a compact subset of some Euclidean space R^p, $p \geq 1$. We also strengthen (A3) to the following.

(A3′) The quantity

$$I\{p(i, j, u, \theta_0) > 0\} \ln(p(i, j, u, \theta)/p(i, j, u, \theta_0))$$

is uniformly bounded in i, j, u, θ and Lipschitz continuous in θ, uniformly with respect to i, j, u.

Both these assumptions will be quite crucial in section 3, and it remains an open problem as to whether one can dispense with them altogether.

As in the preceding chapter, we shall consider three distinct cases: Case 1 (Liapunov condition) and Case 2 (uniform strong recurrence condition) will have the same sets of hypotheses as in section IX.2. We strengthen Case 3 (near-monotonicity condition) by strengthening (A9) to the following.

(A9′) There exists a function $\bar{w} : S \to R^+$ such that

(i) $\lim_{i\to\infty} \bar{w}(i) = \infty$.

(ii) There exist $\bar{a}, \bar{\varepsilon} > 0$ such that for each $\theta, \bar{\theta} \in A$,

$$E^{\theta}[(\bar{w}(X_{n+1}) - \bar{w}(X_n) + \bar{\varepsilon})I\{\bar{w}(X_n) > \bar{a}\}/\mathcal{F}_n] \le 0, \ n \ge 0.$$

under $\gamma\{\xi_{\bar{\theta}}\}$.

(iii) There exists a random variable $\bar{Z}$ and a scalar $\bar{\lambda} > 0$ such that

$$E[\exp(\bar{\lambda}\bar{Z})] < \infty$$

and for any $c \in R^+$, $\theta, \bar{\theta} \in A$,

$$P^{\theta}(|\bar{w}(X_{n+1}) - \bar{w}(X_n)| > c) \le P(\bar{Z} > c), \ n \ge 0,$$

under $\gamma\{\xi_{\bar{\theta}}\}$.

This subsumes (A9). Since we have now replaced θ_0 in (A9) by an arbitrary $\bar{\theta} \in A$, the arguments leading to Theorem IX.2.3 now lead to the stronger conclusion: the map $\theta \to V^{\theta} : A \to R^{\infty}$ is continuous. (Recall that this was true anyway for Cases 1 and 2.) We also have:

Lemma 2.1. (A2) holds.

Proof. This is true in any case for Case 1 and Case 2. For Case 3, argue as for Case 1 using (A9′) and (1.2) to conclude. QED

The appropriate analogue of Lemma IX.1.1 (with $\tilde{\theta}_n, \tilde{L}_n(\cdot)$ replacing $\hat{\theta}_n, L_n(\cdot)$) is also seen to hold.

From now on, we confine ourselves to Case 1. We proceed through a sequence of lemmas. Let $\theta_n \overset{\Delta}{=} \tilde{\theta}_{\sigma(n)}$, $n \ge 1$, for the sake of simplicity. Let $G(i, \theta)$, $i \in S$, $\theta \in A$, denote the set where the map

$$u \to F(\theta, i, u) \overset{\Delta}{=} \sum_{j \in S} p(i, j, u, \theta)V^{\theta}(j) + k(i, u)$$

attains its minimum. Then $G(i, \theta) \subset D$ is compact non-empty for each i, θ. Let

$$G = \bigcup_{i,\theta} \{\theta\} \times \{i\} \times G(i, \theta),$$

viewed as a subset of $A \times S \times D$ with the relative topology.

Lemma 2.2. G is closed.

Proof. Let $(\theta(n), i(n), u(n))$, $n \geq 1$, be a sequence in G converging to some $(\theta, i, \bar{u})$ in $A \times S \times D$. Clearly, $i(n) = i$ from some n on and thus we may assume $i(n) = i$ for all n. By continuity, $F(\theta(n), i, u(n)) \to F(\theta, i, \bar{u})$. For any $u \in D$,

$$F(\theta(n), \; i, u(n)) \leq F(\theta(n), i, u) \to F(\theta, i, u)$$

by our definition of G. Thus $F(\theta, i, \bar{u}) \leq F(\theta, i, u)$ for $u \in D$, implying $(\theta, i, \bar{u}) \in G$. QED

Define the $\mathbb{P}(A \times S \times D)$-valued process $\{\mu_n\}$, $n \geq 1$, by

$$\mu_n(A_1 \times A_2 \times A_3) = \frac{1}{n} \sum_{m=0}^{n-1} I\{\theta_m \in A_1, X_m \in A_2, \xi_m(X_m) \in A_3\}$$

for A_1, A_2, A_3 Borel in A, S, D respectively.

Lemma 2.3. $\{\mu_n\}$ is tight, a.s.

Proof. This is immediate from Lemma 2.1 and the compactness of A, D. QED

Let $\{\bar{\tau}_n\}$ denote the successive return times to state 1.

Lemma 2.4.

$$\sup_{CS,i} E[(\bar{\tau}_{i+1} - \bar{\tau}_i)^n] < \infty \text{ for } n \geq 1,$$

$$\sup_{CS,i} E[(\tau_{i+1} - \tau_i)^n] < \infty \text{ for } n \geq 1.$$

Proof. The first claim is proved by arguments similar to those for Lemma V.3.2. The second follows from the first. QED

Lemma 2.5. Almost surely, there exist $i_0, j_0 \geq 1$ depending on the sample path, such that $\bar{\tau}_{i_0+n} = \tau_{j_0+n}$, $n \geq 1$.

Proof. The sample paths where the above fails lie in the set $\{\bar{\tau}_{i+1} - \bar{\tau}_i > y_i \text{ i.o.}\}$. But

$$\sum_{i=1}^{\infty} P(\bar{\tau}_{i+1} - \bar{\tau}_i > y_i) \leq \sum_{i=1}^{\infty} E[(\bar{\tau}_{i+1} - \bar{\tau}_i)^n]/y_i^n$$

for $n \geq 1$. By the preceding lemma and our choice of $\{y_i\}$, the right hand side is finite for a suitable choice of $n \geq 1$. By the Borel-Contelli lemma,

$$P(\bar{\tau}_{i+1} - \bar{\tau}_i > y_i \text{ i.o.}) = 0$$

and the proof is complete. QED

Lemma 2.6. $\tau_{i+1}/\tau_i \to 1$ a.s., $\bar{\tau}_{i+1}/\bar{\tau}_i \to 1$ a.s.

Proof. We shall prove only the latter, the former then being immediate from the preceding lemma. Using $\sqrt{n}$ in place of y_n in the above proof, it follows that $\bar{\tau}_{n+1} - \bar{\tau}_n$ is evenually dominated by $\sqrt{n}$. On the other hand it exceeds 1 and hence $\bar{\tau}_n$ increases at least as fast as n. The claim follows easily from this. QED

Let C be an open set in A.

Lemma 2.7. Almsot surely, the following holds. If $\{n(k)\}$ is a subsequence of $\{n\}$ such that $\theta_{n(k)} \in C$ and

$$\liminf_{k \to \infty} \frac{1}{n(k)} \sum_{m=1}^{n(k)} I\{\theta_m \in C\} > 0, \tag{2.1}$$

then for all $i \in S$,

$$\liminf_{k \to \infty} \frac{1}{n(k)} \sum_{m=1}^{n(k)} I\{\theta_m \in C,\ X_m = i\} > 0. \tag{2.2}$$

Proof. Given $k \geq 1$, pick $\tau_i \leq n(k) < \tau_{i+1}$. Then

$$\frac{1}{\tau_{i+1}} \sum_{m=1}^{\tau_{i+1}} I\{\theta_m \in C\} - \frac{\tau_{i+1} - \tau_i}{\tau_{i+1}} \leq \frac{1}{n(k)} \sum_{m=1}^{n(k)} I\{\theta_m \in C\}$$

$$\leq \left(\frac{1}{\tau_{i+1}} \sum_{m=1}^{\tau_{i+1}} I\{\theta_m \in C\} \right) \frac{\tau_{i+1}}{\tau_i}. \tag{2.3}$$

Also, for $j \in S$,

$$\frac{1}{\tau_{i+1}} \sum_{m=1}^{\tau_{i+1}} I\{\theta_m \in C,\ X_m = j\} - \frac{\tau_{i+1} - \tau_i}{\tau_{i+1}}$$

$$\leq \frac{1}{n(k)} \sum_{m=1}^{n(k)} I\{\theta_m \in C,\ X_m = j\}$$

$$\leq \left(\frac{1}{\tau_{i+1}} \sum_{m=1}^{\tau_{i+1}} I\{\theta_m \in C,\ X_m = j\} \right) \frac{\tau_{i+1}}{\tau_i}. \tag{2.4}$$

In view of the definition of the KBL scheme and Lemma 2.6, we may replace $\{n(k)\}$ in (2.1) and (2.2) above by $\{\tau_{m(k)+1}\}$ for some subsequence $\{m(k)\}$ of $\{n\}$ satisfying

$$\theta_{\tau_{m(k)}} \in C.$$

Lemma 2.4 allows us to use the martingale stability theorem of [Lo], p. 53, to conclude that

$$\lim_{n\to\infty} \frac{1}{n} \sum_{m=1}^{n} \left[(\tau_{m+1} - \tau_m) - E[(\tau_{m+1} - \tau_m)/\mathcal{F}_{\tau_m}] \right] = 0$$

a.s. for $\mathcal{F}_n = \sigma(X_m, \xi_m,\ m \leq n)$, $n \geq 1$. By Theorem II.3.1 and Lemma 2.4 above, we have

$$1 \leq \liminf_{n\to\infty} \frac{1}{n} \sum_{m=1}^{n} (\tau_{m+1} - \tau_m)$$

$$\leq \limsup_{n\to\infty} \frac{1}{n} \sum_{m=1}^{n} (\tau_{m+1} - \tau_m)$$

$$= \limsup_{n\to\infty} \frac{1}{n} \sum_{m=1}^{n} E[(\tau_{m+1} - \tau_m)/\mathcal{F}_{\tau_m}]$$

$$\leq \sup_{CS} E[\bar{\tau}_2/X_0 = 1] < \infty. \tag{2.5}$$

In view of (2.3)-(2.5) and Lemma 2.6, we may replace (2.1) and (2.2) by

$$\liminf_{k \to \infty} \frac{1}{m(k)} \sum_{m=1}^{m(k)} \sum_{j=\tau_m}^{\tau_{m+1}-1} I\{\theta_j \in C\} > 0 \tag{2.6}$$

and

$$\liminf_{k \to \infty} \frac{1}{m(k)} \sum_{m=1}^{m(k)} \sum_{j=\tau_m}^{\tau_{m+1}-1} I\{\theta_j \in C,\ X_j = i\} > 0. \tag{2.7}$$

Using the martingale stability theorem once more as above, (2.6) and (2.7) respectively a.s. equal

$$\liminf_{k \to \infty} \frac{1}{m(k)} \sum_{m=1}^{m(k)} E\Big[\sum_{j=\tau_m}^{\tau_{m+1}-1} I\{\theta_j \in C\} / \mathcal{F}_{\tau_m} \Big]$$

$$= \liminf_{k \to \infty} \frac{1}{m(k)} \sum_{m=1}^{m(k)} E[(\tau_{m+1} - \tau_m)/\mathcal{F}_{\tau_m}] I\{\theta_{\tau_m} \in C\} \tag{2.8}$$

and

$$\liminf_{k \to \infty} \frac{1}{m(k)} \sum_{m=1}^{m(k)} E\Big[\sum_{j=\tau_m}^{\tau_{m+1}-1} I\{\theta_j \in C,\ X_j = i\} / \mathcal{F}_{\tau_m} \Big]$$

$$= \liminf_{k \to \infty} \frac{1}{m(k)} \sum_{m=1}^{m(k)} E\Big[\sum_{j=\tau_m}^{\tau_{m+1}-1} I\{X_j = i\} / \mathcal{F}_{\tau_m} \Big] I\{\theta_{\tau_m} \in C\}. \tag{2.9}$$

By Lemma 2.4 above and Theorem II.3.1, (2.8) is bounded from above and below by some constant times

$$\liminf_{k \to \infty} \frac{1}{m(k)} \sum_{m=1}^{m(k)} I\{\theta_{\tau_m} \in C\} > 0.$$

Thus by (2.9) the proof is complete if we show that

$$\liminf_{m \to \infty} E\Big[\sum_{j=\tau_m}^{\tau_{m+1}-1} I\{X_j = i\} / \mathcal{F}_{\tau_m} \Big] > 0 \text{ a.s.}$$

By Lemma 2.5, $X_{\tau_m} = 1$ for sufficiently large m. Argue as in the proof of Lemma IX.2.1

and Corollary IX.2.1 to conclude that there exists an integer $N \geq 1$ and a $\rho > 0$ such that under any CS, there exists a path from 1 to i not passing through 1 in between, with probability at least ρ and length at most N. For sufficiently large n, $y_n \geq N$. Thus

$$E\Big[\sum_{j=\tau_m}^{\tau_{m+1}-1} I\{X_j = i\}/\mathcal{F}_{\tau_m}\Big] \geq \rho$$

for sufficiently large m. QED

Let $\{C_n\}$ be an enumeration of open balls in A with rational centres and rational radii less than or equal to, say, 1. From now on, we fix a sample point outside the set of zero probability on which the conclusions of the above lemmas fail, that of Lemma 2.7 being for all $C \in \{C_n, n \geq 1\}$. Let $\bar{\theta}$ be a frequent limit point of $\{\theta_n\}$. For $N \geq 1$, let $B_N \in \{C_n, n \geq 1\}$ be a ball of radius $1/N$ containing $\bar{\theta}$. Fix N for the time being. Let $\{n(k)\} \subset \{n\}$ be such that

$$\theta_{n(k)} = \tilde{\theta}_{n(k)} \in B_N \tag{2.10}$$

and

$$\liminf_{k \to \infty} \frac{1}{n(k)} \sum_{m=1}^{n(k)} I\{\theta_m \in B_N\} > 0. \tag{2.11}$$

That such $\{n(k)\}$ can always be found is proved as follows. Since $\bar{\theta}$ is frequent, we can find $\{m(k)\} \subset \{n\}$ such that (2.11) holds for $\{m(k)\}$ in place of $\{n(k)\}$. One could clearly replace m(k) by τ'_{k+1} where τ'_k = the largest among $\{\tau_i\}$ not exceeding m(k). (Recall the remarks following (2.4)). In turn, this can be replaced by τ'_k, since the difference this makes will be asymptotically vanishing thanks to Lemma 2.6. With $n(k) = \tau'_k$, (2.10) also holds in view of the definition of the KBL scheme.

Let $\bar{\mu}$ be a limit point of $\{\mu_{n(k)}\}$ along, say, $\{n(k_j)\}$. By Lemma 2.7,

$$\bar{\mu}(\bar{B}_N \times \{i\} \times D) > 0, \ i \in S. \tag{2.12}$$

Lemma 2.8. $\bar{\mu}$ is supported on G.

Proof. This is immediate from the fact that $\{\mu_n\}$ are supported on G (by the definition of the KBL scheme) and G is closed. QED

Lemma 2.9. There exist $\theta_i(N)$, $\theta'(N) \in \bar{B}_N$ and $\bar{\xi}_N = [\bar{\xi}_N(1), \bar{\xi}_N(2),...] \in L$ such that for $i \in S$,

$$\bar{\xi}_N(i) \in G(i, \theta_i(N)) \tag{2.13}$$

and

$$p(i, j, \bar{\xi}_N(i), \theta'(N)) = p(i, j, \bar{\xi}_N(i), \theta_0), \ j \in S. \tag{2.14}$$

Proof. Let $\theta'(N)$ be any limit point of $\{\tilde{\theta}_{n(k_j)}\}$. BAs in Corollary IX.1.1, it follows that

$$\sum_{i \in S} \int_{A \times D} \bar{\mu}(dy, \{i\}, du) \Big[\sum_{j \in S} p(i, j, u, \theta_0) \ln(p(i, j, u, \theta'(N))/p(i, j, u, \theta_0)) \Big] = 0. \tag{2.15}$$

Using the strict convexity of $x \to x\ln x$ and Jensen's inequality, familiar arguments (recall section IX.1) show that for any $i \in S$, $\theta' \in A$, $u \in D$,

$$\sum_{j \in S} p(i, j, u, \theta_0) \ln(p(i, j, u, \theta')/p(i, j, u, \theta_0)) \leq 0,$$

with equality if and only if

$$p(i, j, u, \theta') = p(i, j, u, \theta_0), \ j \in S.$$

(2.14) now follows from this and (2.12). (2.13) follows from Lemma 2.8 and (2.12). QED

The main result of this section is the following:

Theorem 2.1. There exists an SS $\gamma\{\bar{\xi}\}$ such that

$$\bar{\xi}(i) \in G(i, \bar{\theta}), \ i \in S, \tag{2.16}$$

and

$$p(i, j, \bar{\xi}(i), \bar{\theta}) = p(i, j, \bar{\xi}(i), \theta_0), \ i, j \in S. \tag{2.17}$$

Proof. As $N \to \infty$, $\theta'(N)$, $\theta_i(N) \to \bar{\theta}$. Let ξ be a limit point of $\{\bar{\xi}_N\}$ in L. (2.16) follows from Lemma 2.2 and (2.17) from (2.14) and the continuity of $p(i, j, ., .)$. QED

Corollary 2.1. $\beta^{\bar{\theta}} \geq \beta^{\theta_0}$.

Proof. (2.16) implies that $\gamma\{\bar{\xi}\}$ is optimal under $\bar{\theta}$, i.e. $\int k d\hat{\pi}^{\bar{\theta}}\{\bar{\xi}\} = \beta^{\bar{\theta}}$. By (2.17),

$$\int k d\hat{\pi}^{\bar{\theta}}\{\bar{\xi}\} = \int k d\hat{\pi}^{\theta_0}\{\bar{\xi}\} \geq \beta^{\theta_0}.$$ QED

X.3 Optimality of the KBL Scheme

This section proves the a.s. optimality of the KBL scheme. As before, we proceed via a long sequence of lemmas.

Recall the definition of $M_n(\theta)$, $n \geq 1$, $\theta \in A$, from section IX.1 and the fact that $M_n(\theta) \leq 0$ for all n, θ. For each θ, $J_n(\theta) = L_n(\theta) - M_n(\theta)$, $n \geq 1$, is an $\{\mathcal{F}_n\}$-martingale with zero mean and bounded increments (cf. (A3)). Denote by $\langle J(\theta)\rangle_n$, $n \geq 1$, its quadratic variation process. Let $\delta \in (1/2, \alpha)$.

Lemma 3.1. For each $\theta \in A$,

$$\lim_{n \to \infty} n^{-\delta} J_n(\theta) = 0 \text{ a.s.} \tag{3.1}$$

Proof. Let $\langle J(\theta)\rangle_\infty = \lim_{n \to \infty} \langle J(\theta)\rangle_n \in R^+ \cup \{\infty\}$. By Proposition VIII.2.3(c) of [Nev], $J_n(\theta)$ converges a.s. on $\{\langle J(\theta)\rangle_\infty < \infty\}$. By Proposition VII.2.7 of [Nev],

$$\limsup_{n \to \infty} \frac{|J_n(\theta)|}{\sqrt{2\langle J(\theta)\rangle_n \ln(\ln\langle J(\theta)\rangle_n)}} = 1 \text{ a.s.}$$

on $\{\langle J(\theta)\rangle_\infty = \infty\}$. Since $\{\langle J(\theta)\rangle_n\}$ has bounded increments (because $\{J_n(\theta)\}$ does), it grows at most linearly with n. The claim is immediate from this and the foregoing. QED

Lemma 3.2.

$$\lim_{n \to \infty} \sup_{\theta \in A} n^{-\delta} J_n(\theta) = 0 \text{ a.s.} \tag{3.2}$$

Proof. Outside a set N of zero probability, (3.1) holds for all θ in a countable dense set in A. The claim will follow from this if we show that the map $\theta \to n^{-\delta} J_n(\theta)$ is continuous uniformly in n, a.s. Let $\{Y_n(\theta)\}$ denote the increments of $\{J_n(\theta)\}$. Let $\theta_1, \theta_2 \in A$ and define $\{Z_n\}$ by

$$Z_n = \begin{cases} (J_n(\theta_1) - J_n(\theta_2))N^{-\delta}, & 1 \leq n \leq N, \\ Z_N, & n > N, \end{cases}$$

for some $N \geq 1$. Then $\{Z_n\}$ is an $\{\mathcal{F}_n\}$-martingale. Applying Corollary 1, p. 385 of [ChTe], we get

$$E[\,|N^{-\delta}(J_N(\theta_1) - J_N(\theta_2))|^{2k}] \le KE\,[N^{-2\delta k}\,(\sum_{n=1}^{N} |Y_n(\theta_1) - Y_n(\theta_2)|^2)^k]$$

for $k \ge 1$ with K = a suitable constant independent of N but depending on k. Let $\varepsilon = 2\delta - 1 > 0$. By (A3′), the right-hand side can be bounded by

$$K'N^{-\varepsilon k}\,\|\theta_1 - \theta_2\|^{2k} \tag{3.3}$$

for a suitable constant K' independent of N, but depending on K above and the Lipschitz constant in (A3′). Recall that A is a closed bounded set in some Euclidean space R^p, $p \ge 1$. We assume that A is the unit cube in R^p. The proof can be easily adapted to the general case. Let Q be the set of p-vectors with elements in $\{0,1\}$. For $n = 1,2,\ldots$, let Q'_n be the set of p-vectors with elements in $\{0, 1,\ldots,2^n\}$. Let $Q_n = \{\bar{k}/2^n \,|\, \bar{k} \in Q'_n\}$ and $\bar{Q} = \bigcup_n Q_n$. For $N \ge 1$, $n \ge 1$, let

$$Z'_{nN} = \sup_{\bar{k} \in Q'_n} \max_{\ell \in Q} \sup_{m \le N} |m^{-\delta}\Big(J_m\Big(\frac{\bar{k}+\bar{\ell}}{2^n}\Big) - J_m\Big(\frac{\bar{k}}{2^n}\Big)\Big)|\,.$$

If $\|\theta_1 - \theta_2\| < 2^{-n}$ and $\theta_1, \theta_2 \in \bar{Q}$, then there exists a $\bar{k} \in Q'_n$ such that $\|\theta_i - \bar{k}/2^n\| < 2^{-n}$ for $i = 1,2$. Thus for $1 \le i \le p$, the i'th component of θ_1, θ_2 must be of the form

$$\theta_j(i) = \frac{k_i}{2^n} \pm \sum_{\ell=1}^{m_j(i)} t_j(\ell)2^{-(n+\ell)},\ j = 1 \text{ or } 2,$$

where $\bar{k} = [k_1,\ldots,k_p]^T$ and $t_j(\ell) = 0$ or 1. Thus for $i = 1$ or 2,

$$\sup_{m \le N} |m^{-\delta}\Big(J_m(\theta_i) - J_m\Big(\frac{\bar{k}}{2^n}\Big)\Big)| \le \sum_{j=n+1}^{\infty} Z'_{jN}\,,$$

leading to

$$\sup_{m \le N} |m^{-\delta}(J_m(\theta_1) - J_m(\theta_2))| \le 2 \sum_{j=n+1}^{\infty} Z'_{jN}.$$

Let $\bar{\varepsilon}, \eta > 0$. Then

$$P\Big(\sup_{\substack{\theta_1,\theta_2 \in \bar{Q} \\ \|\theta_1-\theta_2\| \le 2^{-n}}} \sup_{m \le N} (m^{-\delta} | J_m(\theta_1) - J_m(\theta_2) |) \ge \bar{\varepsilon} \Big)$$

$$\le P\Big(\sum_{j=n+1}^{\infty} Z'_{jN} \ge \bar{\varepsilon}/2 \Big). \qquad (3.4)$$

Pick $k \ge 1$ in (3.3) such that $\varepsilon k > 1$ and $2k > p + 1$. Let $0 < r < (2k-p)/2k$ and set $s = 2k - p - 2rk$. Then $s > 0$. Pick n such that

$$\sum_{j=n+1}^{\infty} \frac{1}{2^{jr}} < \frac{\bar{\varepsilon}}{2}.$$

Then

$$P(\sum_{j=n+1}^{\infty} Z'_{jN} > \bar{\varepsilon}/2) \le P(\sum_{j=n+1}^{\infty} Z'_{jN} > \sum_{j=n+1}^{\infty} \frac{1}{2^{jr}})$$

$$\le \sum_{j=n+1}^{\infty} P(Z'_{jN} > \frac{1}{2^{jr}}).$$

But

$$P(Z'_{jN} > \frac{1}{2^{jr}}) \le \sum_{\bar{k} \in Q'_j} \sum_{\bar{\ell} \in Q} P(\sup_{m \le N} (m^{-\delta} | J_m\Big(\frac{\bar{k}+\bar{\ell}}{2^j}\Big) - J_m\Big(\frac{\bar{k}}{2^j}\Big) |) \ge \frac{1}{2^{jr}}) \qquad (3.5)$$

and for $\theta, \theta' \in A$,

$$P(\sup_{m \le N} | m^{-\delta}(J_m(\theta) - J_m(\theta')) | \ge \frac{1}{2^{jr}})$$

$$\le \sum_{n=1}^{N} P((n^{-\delta} | J_n(\theta) - J_n(\theta') |)^{2k} \ge (\frac{1}{2^j})^{2kr})$$

$$\le 2^{2kjr} \sum_{m=1}^{N} K' m^{-\varepsilon k} \| \theta - \theta' \|^{2k}. \qquad (3.6)$$

Since $\| (\bar{k}+\bar{\ell})/2^j - \bar{k}/2^j \| \le \sqrt{p}/2^j$ for $\bar{k} \in Q'_j$ and $\bar{\ell} \in Q$, we have from (3.5) and (3.6) that for suitably defined K'',

$$p(Z'_{jN} > 2^{-jr}) \le K'.2^p.p^k.2^{2kjr}.2^{pj}.\frac{1}{2^{2kj}}\left(\sum_{m=1}^{\infty} m^{-\varepsilon k}\right)$$

$$= K'' 2^{-js}.$$

Pick n large enough so that

$$\sum_{j=n+1}^{\infty} \frac{1}{2^{js}} < \frac{\eta}{K_0}$$

where $K_0 = K'.2^p.p^k.\left(\sum_{i=1}^{\infty} i^{-\varepsilon k}\right)$. Then

$$P\left(\sum_{j=n+1}^{\infty} Z'_{jN} > \frac{\bar{\varepsilon}}{2}\right) < \eta.$$

By (3.4),

$$P\left(\sup_{\substack{\theta_1,\theta_2 \in \bar{Q} \\ \|\theta_1-\theta_2\| \le 2^{-n}}} \sup_{m \le N} (m^{-\delta} | J_m(\theta_1) - J_m(\theta_2) |) \ge \bar{\varepsilon}\right) < \eta .$$

Letting $N \to \infty$, one gets

$$P\left(\sup_{\substack{\theta_1,\theta_2 \in \bar{Q} \\ \|\theta_1-\theta_2\| \le 2^{-n}}} \sup_{m < \infty} (m^{-\delta} | J_m(\theta_1) - J_m(\theta_2) |) \ge \bar{\varepsilon}\right) < \eta .$$

Standard arguments now show that the map $\theta \to n^{-\delta} J_n(\theta)$ has a continuous version, continuous uniformly in n, a.s. This completes the proof. QED

Recall the set-up of the preceding section in which Theorem 2.1 was proved. Enlarge the set of zero probability left out there to include the set where the statement of the above lemma fails. Let $\bar{\theta}$ etc. be as before.

Lemma 3.3. $\beta^{\bar{\theta}} = \beta^{\theta_0}$.

Proof. Let $\{n(k)\} \subset \{n\}$ be such that $\tilde{\theta}_{n(k)} \to \bar{\theta}$. In view of the non-positivity of $M_n(\theta)$ above,

$$0 = \tilde{L}_{n(k)}(\theta_0) \le \tilde{L}_{n(k)}(\tilde{\theta}_{n(k)})$$

$$\le -n(k)^{\alpha} \ln(\beta^{\tilde{\theta}_{n(k)}}/\beta^{\theta_0}) + n(k)^{\delta} \sup_{\theta} (n(k)^{-\delta} J_{n(k)}(\theta)). \tag{3.6}$$

If $\theta(n) \to \theta$ in A and $\gamma\{\bar{\xi}_n\}$ are optimal SSS for $\theta(n)$ respectively, then $(\theta(n), i, \bar{\xi}_n(i)) \in G$ for all i, n and hence $(\theta, i, \bar{\xi}(i)) \in G$ for all i and for all limit points $\bar{\xi}$ of $\{\bar{\xi}_n\}$ in L. That is,

$$k(i, \bar{\xi}(i)) + \sum_j p(i, j, \bar{\xi}(i), \theta)\, V^{\theta}(j) = \min_u (k(i, u) + \sum_j p(i, j, u, \theta)\, V^{\theta}(j)), \ i \ge 1.$$

Thus $\gamma\{\bar{\xi}\}$ must be an optimal SSS under θ. By dropping to the appropriate subsequence, let $\bar{\xi}_n \to \bar{\xi}$. Now

$$\pi^{\theta(n)}\{\bar{\xi}_n\} = \pi^{\theta(n)}\{\bar{\xi}_n\}P^{\theta(n)}\{\bar{\xi}_n\}, \ n = 1,2,\ldots$$

Under our hypotheses, $\pi^{\theta(n)}\{\bar{\xi}_n\}$, $n \ge 1$, are tight. Let π be a limit point thereof in $\mathbb{P}(S)$ and hence by Scheffe's theorem, in total variation. Passing to the limit as $n \to \infty$ along an appropriate subsequence in the above equation, familiar arguments yield

$$\pi P^{\theta}\{\bar{\xi}\} = \pi,$$

implying $\pi = \pi^{\theta}\{\bar{\xi}\}$. It follows that

$$\beta^{\theta(n)} = \int k d\hat{\pi}^{\theta(n)}\{\bar{\xi}_n\} \to \int k d\hat{\pi}^{\theta}\{\bar{\xi}\} = \beta^{\theta}.$$

Thus the map $\theta \to \beta^{\theta}$ is continuous. Therefore

$$\beta^{\tilde{\theta}_{n(k)}} \to \beta^{\bar{\theta}}.$$

If $\beta^{\bar{\theta}} > \beta^{\theta_0}$, $\beta^{\tilde{\theta}_{n(k)}}$ is bounded away from β^{θ_0} for sufficiently large k. Use this in (3.6) along with the fact that $\alpha > \delta$ and the preceding lemma, to conclude that the right hand side of (3.6) becomes strictly negative for large k, a contradiction. Thus $\beta^{\bar{\theta}} \leq \beta^{\theta_0}$. The claim now follows from Corollary 2.1. QED

Corollary 3.1. $V^{\bar{\theta}} = V^{\theta_0}$ and $G(i, \bar{\theta}) = G(i, \theta_0)$, $i \in S$.

Proof. With $k\{\xi\} = [k(1, \xi(1)), k(2, \xi(2)),...]^T$ for $\xi \in L$, we have

$$\beta^{\theta_0} 1_c = (P^{\bar{\theta}} \{\xi_{\theta}\} - I) V^{\bar{\theta}} + k\{\xi_{\bar{\theta}}\},$$

where we have used the above lemma and the optimality of ξ_θ under $\bar{\theta}$. Since

$$P^{\bar{\theta}}\{\xi_\theta\} = P^{\theta_0}\{\xi_{\bar{\theta}}\}$$

by Theorem 2.1, we have

$$\beta^{\theta_0} 1_c = (P^{\theta_0} \{\xi_{\bar{\theta}}\} - I) V^{\bar{\theta}} + k\{\xi_{\bar{\theta}}\}. \tag{3.7}$$

On the other hand, the dynamic programming equations yield

$$\beta^{\theta_0} 1_c \leq (P^{\theta_0} \{\xi_{\bar{\theta}}\} - I) V^{\theta_0} + k\{\xi_{\bar{\theta}}\}. \tag{3.8}$$

Let $\{X_n\}$ be governed by $\gamma\{\xi_{\bar{\theta}}\}$ with initial law $\pi^{\theta_0}\{\xi_{\bar{\theta}}\}$. From (3.7), (3.8), it follows that $V^{\theta_0}(X_n) - V^{\bar{\theta}}(X_n)$ is an $\{\mathcal{F}_n\}$-submartingale. But

$$\sup_n E[\,|\,V^{\theta_0}(X_n) - V^{\bar{\theta}}(X_n)\,|\,] \leq C_1 + C_2 \int w d\pi^{\theta_0}\{\xi_{\bar{\theta}}\} < \infty$$

for suitable constants $C_1, C_2 > 0$ (cf. section VI.4). Thus the submartingale convergence theorem applies and $V^{\theta_0}(X_n) - V^{\bar{\theta}}(X_n)$ converges a.s. as $n \to \infty$. By positive recurrence, this is possible only if $V^{\theta_0}(i) - V^{\bar{\theta}}(i)$ does not depend on i. But then it must be zero since it is zero for $i = 1$. This proves the first claim. The second claim follows immediately from the first. QED

Theorem 3.1. The KBL scheme is a.s. optimal.

Proof. To start with, consider a fixed sample path as above. Let μ^* be a limit point of

$\{\mu_n\}$. It is clear that each θ in the support of the image of μ^* under the projection $A \times S \times D \to A$ is a frequent limit point of $\{\theta_n\}$. By Lemma V.1.1, the image of μ^* under the projection $A \times S \times D \to S \times D$ is of the form $\hat{\pi}^{\theta_0}[\Phi]$ for some SSRS $\gamma[\Phi]$, $\Phi = \prod_i \hat{\Phi}_i$. From the preceding lemma and the fact that μ^* is supported on G, it follows that $\hat{\Phi}_i$ is supported on $G(i, \theta_0)$ for each $i \in S$. Hence $\gamma[\Phi]$ is an optimal SSRS, that is

$$\int k d\mu^* = \beta^{\theta_0}.$$

The claim follows. QED

The proof for Case 2 is similar. As for Case 3, our definition of the KBL scheme and (A9′) all but reduce it to Case 1.

We conclude with a remark concerning our choice of the sequence $\{y_i\}$. Note that the foregoing would work equally well if $\{\bar{\tau}_i\}$ instead of $\{\tau_i\}$ were the times at which one updated the control policy. The insertion of y_i's as above is a practical necessity and not a theoretical need. The return times to a single state may not occur fast enough, spoiling the transient response of the KBL scheme. On the other hand, one can keep the y_i's small for small i's, facilitating frequent updating of controls, hopefully bringing the system into a satisfactory regime of operation in a short time. High values of y_i for large i then allow further discrimination between available models by testing each candidate for a sufficiently long time. This leads to an asymptotically optimal behaviour. Thus the y_i's may be considered additional, deterministic decision variables. A rigorous study as to what would be the best (or a 'good') choice is lacking. See the next chapter for further comments on this.

Chapter XI
Concluding Remarks

This chapter concludes with a brief list of some allied developments and a list of some unsolved problems suggested by the material covered in this text.

XI.1 Some Related Developments

As a pointer to where things are heading, here is a very short and non-exhaustive list of some recent related developments . A representative reference is provided for each.

(1) Adaptive filtering of Markov chains [ArMa] and adaptive control of partially observed Markov chains [HeMa].

(2) Decentralized control [HsMa] and multilayer control [FoVa].

(3) Singular perturbation analysis of controlled Markov chains with rare transitions [DeQ].

(4) Computational methods [Tijm] and implementational aspects [MaSc].

(5) Alternative approaches to adaptive control, involving alternative cost crtieria [Scha] or alternative estimation schemes [VaH].

(6) Applications to specific problems where additional structure is available, such as bandit problems [BVW] or controlled networks of queues {RVW].

(7) Controlled Markov chains on more general state spaces [BhMa].

XI.2 Open Problems

(1) In the framework of Chapter V, prove or disprove the following.

(i) If all SS are SSS, 'Condition A' will automatically hold.

(ii) If the stable case, 'Condition B' is not necessary for the existence of an SS optimal among all CS. 'Condition A' suffices.

(2) A related issue is that of structural stability for Markov chains: how do its recurrence

properties (transience/null or positive recurrence) alter under perturbations of the transition matrix?

(3) In the framework of Chapter VI, give necessary and sufficient conditions under which local optimality implies optimality.

(4) The conditions under which we have carried out the existence/dynamic programming programme for ergodic control in Chapterv V and VI do not exhaust the possibilities. Find other conditions that allow the programme to be carried through.

(5) The same for the programme of Chapters IX and X for adaptive control.

(6) Extend our results for the constrained control problem to the near-monotone case.

(7) Develop computational schemes for finding the optimal policy in the constrained control problem.

(8) The convex analytic approach of Chapters II-IV has been only partially extended to the partially observed case in Chapter VIII. In particular, no direct argument is available for the extremality of occupation measures corresponding to SS (or MS, as the case may be). Complete this programme.

In this connection, it should be noted that the convex analytic approach for the partially observed case is much less economical than the conventional approach. It would, however, have one advantage over the latter if carried through: the results of Chapter VII for constrained control problems would go over easily to the partially observed case.

(9) Extend the existence/dynamic programming results for ergodic control under complete observations to the partially observed case.

(10) Study the self-tuning scheme and the KBL scheme in the case when the prescribed parametrized family of models does not include the true model. See [BoV3] for some partial developments along these lines.

(11) Prove or disprove the conjecture: the costs (C4) and (C4′) (or (C4″)) of Chapter VIII are a.s. equal under any CS admissible in the sense of Chapter VIII.

(12) For adaptive control in the near-monotone case, we have assumed Liapunov-type stability conditions under special combinations of parameters and controls. Intuitively, near-monotonicity discourages instability and hence introduces a 'centripetal' behaviour in the optimally controlled Markov chain. Near-monotonicity that is uniform in the unknown parameter should induce such behaviour uniformly with respect to the unknown parameter. Thus stability conditions of the type mentioned above should be a consequence of near-

monotonicity, and need not be assumed separately. Confirm this intuition.

(13) Recall the sequence $\{y_i\}$ of Chapter X which determines the times at which we update the control policy in the KBL scheme. It was a determinsitic sequence. More generally,we can allow it to be a sequence of stopping times which can then be treated as additional decision variables, introducing further adaptive behaviour in the scheme. Explore this possibility.

(14) Extend the results of Chapter X to more general parameter spaces.

Added in proof: While this book was in press, (6) above has been settled by the author. See [Bor 10] for details.

Appendix: Spaces of Probability Measures

This is a largely self-contained account of spaces of probability measures on Polish spaces with Prohorov topology (also called the topology of weak convergence) and some related topics. The principal reference is [Bill].

A.1 Polish Spaces

Recall that a Polish space is a topological space which is separable and admits a complete metrization. Examples of such spaces are: separable Banach spaces, compact metric spaces, the space $D[0,1]$ of cadlag paths from $[0,1]$ to R with Skorohod topology [Bill] and, as we shall see later, spaces of probability measures on Polish spaces with the Prohorov topology.

This section proves some results about Polish spaces that will be used later. Let S be a Polish space and $d(.,.)$ a complete metric on it. By replacing $d(.,.)$ by $d(.,.)/(1 + d(.,.))$ if necessary, we may assume that $d(.,.)$ takes values in $[0,1)$. We endow S with its Borel σ-field.

Theorem A.1.1. S is homeomorphic to a G_δ subset of $[0,1]^\infty$.

Proof. Let $\{s_n\} \subset S$ be dense. Define $h : S \to [0,1]^\infty$ by $h(x) = [d(x, s_1), d(x, s_2),\ldots]$. If $x_n \to x$ in S, $h(x_n) \to h(x)$ in $[0,1]^\infty$. Conversely, if $x_n \not\longrightarrow x$ in S, $d(x_n, x) > \varepsilon$ infinitely often for some $\varepsilon > 0$, and hence there is an s_k satisfying $d(x_n, s_k) > \varepsilon/2 \geq (d(x, s_k)$ for infinitely many n. Thus $h(x_n) \not\longrightarrow h(x)$ in $[0, 1]^\infty$. This proves that h is a homeomorphism. Metrize the product topology of $[0, 1]^\infty$ with some metric $\bar{d}(.,.)$, say

$$\bar{d}((x_1, x_2,\ldots), (y_1, y_2,\ldots)) = \sum_{n=1}^{\infty} 2^{-n} | x_n - y_n |.$$

For each $h(x)$ with $x \in S$, take B open in $[0,1]^\infty$ such that $h(x) \in B$ and the diameters of B, $h^{-1}(B)$ are less than $1/m$ for some prescribed m. Let G_m be the union of all such B. Then G_m is open and $h(S) \subset \bigcap_m G_m$. Let $y \in \bigcap_m G_m$. For each m, there exists an open ball U_m containing y with a centre at $h(x_m)$ for some $x_m \in S$, such that the diameters of U_m and $V_m = h^{-1}(U_m)$ are less than $1/m$. Since $h(x_m) \to y$, $y \in \overline{h(S)}$. For any $k, m \geq 1$, the neighbourhood $U_k \cap V_m$ of y contains a $\bar{y} \in h(S)$. Then $\bar{x} = h^{-1}(\bar{y})$ is in $V_k \cap V_m$ and

$$d(x_k, x_m) \le d(x_k, \bar{x}) + d(\bar{x}, x_m) < \frac{1}{k} + \frac{1}{m} .$$

Thus $\{x_m\}$ is Cauchy and has a limit x. Then $h(x) = y$. Thus $y \in h(S)$, proving that $h(S) = \bigcap_m G_m$ which is a G_δ set. QED

Since $[0, 1]^\infty$ is compact, $\overline{h(S)}$ above is also compact and S is densely homeomorphically embedded in $\overline{h(S)}$. We call $\overline{h(S)}$ the compactification of S.

Theorem A.1.2. S is isomorphic (as a measurable space) to a measurable subset of [0,1].

Proof. In view of the preceding theorem, it suffices to consider $S = [0, 1)^\infty$. We shall prove that $[0,1)^\infty$ is isomorphic to [0,1). Construct $f : (0,1) \to [0,1)^\infty$ as follows. Write $x \in [0,1)$ as $x = 0.\, x_1\, x_2\, x_3\, \ldots$, its unique binary expansion containing infinitely many zeroes. Consider the array

1			
2	3		
4	5	6	
7	8	9	10

and so on. Let $f(x) = (f_1(x), f_2(x),\ldots)$ with $f_1(x) = 0.x_1x_2x_4x_7\ \ldots$, $f_2(x) = 0.x_3x_5x_8\ \ldots$, $f_3(x) = 0.x_6x_9\ \ldots$ etc. (The subscripts are obtained by going down the corresponding column of the above array.) Conversely, let $x = [x^1, x^2,\ldots] \in [0, 1)^\infty$ with $x^k = 0.x_1^k\, x_2^k \ldots$ the corresponding unique binary expansions with infinitely many zeros. Let g(x) = the binary decimal whose n-th entry is x_j^k, if n appears in the k-th column j numbers down. Then f, g are measurable and $f = g^{-1}$. QED

A probability measure P on S is said to be tight if for each $\varepsilon > 0$, there exists a compact set $K_\varepsilon \subset S$ with $P(K_\varepsilon) \ge 1 - \varepsilon$. Analogously, a family P_α, $\alpha \in I$, (I being an index set) is said to be tight if the above holds for all P_α uniformly in α, i.e. the set K_ε above can be chosen to be the same for all α.

Theorem A.1.3 (Oxtoby, Ulam) Every probability measure on a Polish space is tight.

Proof. For each $n \geq 1$, let $\{A_{ni}\}$ be a collection of open balls of diameters not exceeding $1/n$ and covering S. Let $\varepsilon > 0$. Pick $n_i \geq 1$, $i = 1,2,\ldots$, such that

$$P(\bigcup_{j \leq n_i} A_{ij}) > 1 - \varepsilon 2^{-i}.$$

Then the set

$$\bigcap_{i \geq 1} \bigcup_{j \leq n_i} A_{ij} \subset S$$

is totally bounded and has a compact closure K_ε with

$$P(K_\varepsilon) \geq 1 - \varepsilon .$$

QED

Theorem A.1.4. A family of probability measures on a finite or countably infinite product of Polish spaces is tight if and only if its image measures under the projection onto each of the factor spaces form a tight family.

Proof. Let $I = \{1,2,\ldots\}$ be a finite or countable set and S_α, $\alpha \in I$, a collection of Polish spaces with complete metrics d_α, $\alpha \in I$, respectively. Let $S' = \prod_{\alpha \in I} S_\alpha$ with the product σ-field. Let P_β, $\beta \in J$, be a tight family of probability measures on S'. Then for each $\varepsilon > 0$, there exists a compact set $K \subset S$ such that $P_\beta(K) > 1 - \varepsilon$ for each $\beta \in J$. Letting $p_\alpha : S \to S_\alpha$ denote the projection map for $\alpha \in I$, let $\mu_{\beta\alpha}$ denote the image of P_β under p_α. Let K_α denote the image of K under p_α. Then K_α is compact and

$$\mu_{\beta\alpha}(K_\alpha) \geq P_\beta(K) > 1 - \varepsilon \text{ for all } \beta \in J.$$

This proves the 'only if' part. Conversely, suppose that for each $\alpha \in I$, $\{\mu_{\beta\alpha}, \alpha \in I\}$ is a tight family. For $\alpha = 1,2,\ldots$, let $K_\alpha \subset S_\alpha$ be a compact set such that

$$\mu_{\beta\alpha}(K_\alpha) > 1 - \varepsilon 2^{-\alpha}, \ \beta \in J.$$

Then $K = \Pi_\alpha K_\alpha$ is compact in S' with $P_\beta(K) > 1 - \varepsilon$ for all $\beta \in J$. This proves the 'if' part.

QED

A.2 The Prohorov Topology

Let $C_b(S)$, $\mathbb{P}(S)$ denote respectively the space of bounded continuous real valued functions on S, and the space of probability measures on S. Endow $C_b(S)$ with the supremum norm $\|\cdot\|$. $\mathbb{P}(S)$ will be given the topology for which a local base at $P \in \mathbb{P}(S)$ is given by the sets of the type

(i) $$\{Q \in \mathbb{P}(S) \mid |\int f_i\, dQ - \int f_i\, dP| < \varepsilon_i,\ 1 \le i \le k\}$$

for some $k \ge 1$, $\varepsilon_i > 0$ and $f_i \in C_b(S)$ for $1 \le i \le k$.

It is easily seen that this topology is Hausdorff and is coarser than the one induced by the total variation norm. It is called the Prohorov topology or the topology of weak convergence. Some other possible choices for a local basis at $P \in \mathbb{P}(S)$ are given below. (Here, $\varepsilon_i > 0$, $k \ge 1$ are arbitrary in (ii)-(v)).

(ii) $$\{Q \in \mathbb{P}(S) \mid Q(F_i) < P(F_i) + \varepsilon_i,\ 1 \le i \le k\},\ F_i \subset S \text{ closed.}$$

(iii) $$\{Q \in \mathbb{P}(S) \mid Q(G_i) > P(G_i) - \varepsilon_i,\ 1 \le i \le k\},\ G_i \subset S \text{ open,}$$

(iv) $$\{Q \in \mathbb{P}(S) \mid |Q(A_i) - P(A_i)| < \varepsilon_i,\ 1 \le i \le k\},\ A_i \subset S$$

satisfy $P(\partial A_i) = 0$ where ∂A_i is the boundary of A_i,

(v) $$\{Q \in \mathbb{P}(S) \mid |\int f_i\, dP - \int f_i\, dQ| < \varepsilon_i,\ 1 \le i \le k\},$$

f_i are bounded and uniformly continuous with respect to some metric ρ on S which is equivalent to d.

Theorem A.2.1. The above local bases are equivalent.

Proof. Since (ii), (iii) coincide for $G_i = F_i^c$, we need consider only (ii). Fix $P \in \mathbb{P}(S)$.

Step 1: A set of type (iv) contains a set of type (ii).

Let H be a set of the type described in (iv). Then each Q in a set of type (ii), with $\{F_i, 1 \le i \le k\}$ replaced by $\{\bar{A}_i, \bar{A}_i^c, 1 \le i \le k\}$ and ε_i by $\varepsilon = \min_i \varepsilon_i$ satisfies (with $\mathring{B}$ denoting the interior of a set $B \subset S$),

$$Q(A_i) \le Q(\bar{A}_i) < P(\bar{A}_i) + \varepsilon = P(A_i) + \varepsilon,$$

$$Q(A_i) \ge Q(\mathring{A}_i) > P(\mathring{A}_i) - \varepsilon = P(A_i) - \varepsilon,$$

for $1 \le i \le k$. Thus $|Q(A_i) - P(A_i)| < \varepsilon$, $1 \le i \le k$, implying that $Q \in H$.

Step 2: A set of type (ii) contains a set of type (iv).

Let H be a set of type (ii). One can find a $\delta > 0$ such that $F_i^{\delta} = \{x \mid d(x, F_i) < \delta\}$ satisfy $P(\partial F_i) = 0$ and $P(F_i^{\delta}) < P(F_i) + \varepsilon_i/2$ for all i. If $|Q(F_i^{\delta}) - P(F_i^{\delta})| < \varepsilon_i/2$ for all i, then $Q(F_i) < P(F_i) + \varepsilon_i$ for all i. Thus a set of type (ii) contains a set of type (iv).

Step 3: A set of type (ii) contains a set of type (i) or (v).

Let H, F_i, δ be as above. Define $\varphi : R \to R$ by

$$\varphi(t) = \begin{cases} 1 & \text{if } t \le 0 \\ 1 - t & \text{if } 0 \le t \le 1 \\ 0 & \text{if } t \ge 1 \end{cases}$$

and set $f_i(x) = \varphi(\delta^{-1} \rho(x, F_i))$, $1 \le i \le k$. Then f_i are uniformly continuous with respect to ρ. For each i, $f_i(x) \in [0,1]$ for all x, $f_i(x) = 0$ on $(F_i^{\delta})^c$ (Here F^{δ} is defined as before but with ρ replacing d) and $f_i(x) = 1$ on F_i. If

$$|\int f_i \, dP - \int f_i dQ| < \varepsilon_i/2, \ 1 \le i \le k,$$

then for $1 \le i \le k$,

$$\begin{aligned} Q(F_i) &\le \int f_i \, dQ \le \int f_i \, dP + \varepsilon_i/2 \\ &\le P(F_i^{\delta}) + \varepsilon_i/2 \\ &< P(F_i) + \varepsilon_i . \end{aligned}$$

Thus H contains a set of type (v) and hence of type (i).

Step 4: A set of type (i) or (v) contains a set of type (ii).

Consider a set of type (i). By adding a constant to each f_i and scaling it if necesary, we may assume that $f_i(S) \subset (0,1)$ for $1 \le i \le k$. Let $\varepsilon = \min_i \varepsilon_i$, and pick $m \ge 1$ such that $m^{-1} < \varepsilon/2$. Set $F_{ij} = \{x \mid j/m \le f_i(x)\}$ for $0 \le j \le m$, $1 \le i \le k$. For any $\tilde{Q} \in \mathbb{P}(S)$, we have

$$\sum_{j=1}^{m} \frac{j-1}{m} \tilde{Q}\left(\frac{j-1}{m} \le f_i(x) < \frac{j}{m}\right) \le \int f_i \, d\tilde{Q}$$

$$\leq \sum_{j=1}^{m} \frac{j}{m} \tilde{Q}\left(\frac{j-1}{m} \leq f_i(x) < \frac{j}{m}\right).$$

The sum on the right is

$$\sum_{j=1}^{m} \frac{j}{m} (\tilde{Q}(F_{i(j-1)}) - \tilde{Q}(F_{ij}))$$

$$= \frac{1}{m} + \frac{1}{m} \sum_{j=1}^{m} \tilde{Q}(F_{ij})$$

and the sum on the left is

$$\sum_{j=1}^{m} \frac{j-1}{m} (\tilde{Q}(F_{i(j-1)}) - \tilde{Q}(F_{ij}))$$

$$= \frac{1}{m} \sum_{j=1}^{m} \tilde{Q}(F_{ij}).$$

Letting $\tilde{Q} = Q$, we get

$$\int f_i \, dQ < \frac{\varepsilon_i}{2} + \frac{1}{m} \sum_{j=1}^{m} Q(F_{ij}).$$

Letting $\tilde{Q} = P$, we get

$$\frac{1}{m} \sum_{j=1}^{m} P(F_{ij}) < \int f_i \, dP.$$

Thus if Q satisfies: $Q(F_{ij}) \leq P(F_{ij}) + \varepsilon_i/2$, then $\int f_i \, dQ < \int f_i \, dP + \varepsilon_i$. Applying the same argument to $1 - f_i$ in place of f_i, we conclude that a set of type (i) or (v) contains a set of type (ii).

This completes the proof. QED

Corollary A.2.1. $P_n \to P$ in $\mathbb{P}(S)$ if and only if any one of the following holds.

(1) $\int f dP_n \to \int f dP$ for $f \in C_b(S)$.

(2) $\limsup_{n \to \infty} P_n(F) \leq P(F)$ for all closed $F \subset S$.

(3) $\liminf_{n \to \infty} P_n(G) \geq P(G)$ for all open $G \subset S$.

(4) $\lim_{n \to \infty} P_n(A) = P(A)$ for all $A \subset S$ with $P(\partial A) = 0$.

(5) $\int fdP_n \to \int fdP$ for all $f : S \to R$ which are bounded and uniformly continuous with respect to some metric ρ on S which is equivalent to d.

This is immediate from the above theorem.

Corollary A.2.2. $\mathbb{P}(S)$ is separable.

Proof. Consider a set H of type (ii) above. For every nonempty set B_j in the finite partition generated by the collection $\{F_i,\ 1 \leq i \leq k\}$, take a point $b(j) \in B_j$. Then the probability measure $\sum_j P(B_j)\, \delta_{b(j)}$ (where δ_x dentoes the Dirac measure at $x \in S$) lies in H. Thus probability measures with finite supports are dense in $\mathbb{P}(S)$. Hence the countable set of probability measures with finite supports lying in a prescribed countable dense subset of S and having a rational mass at each point of the support is dense in $\mathbb{P}(S)$. QED

Call $D \subset C_b(S)$ a convergence determining class if $\int fdP_n \to \int fdP$ for $f \in D$ and $\{P_n\}$, P in $\mathbb{P}(S)$ implies $P_n \to P$ in $\mathbb{P}(S)$. Clearly, $C_b(S)$ itself is a convergence determining class.

Theorem A.2.2. There exists a countable convergence determining class.

Proof. Let $h, \bar{d}$ be as before. Then $d'(x, y) = \bar{d}(h(x), h(y))$ for $x, y \in S$ is a metric on S which is equivalent to d. If $f \in C_b(S)$ is uniformly continuous with respect to d', $\bar{f} = f \circ h^{-1} \in C_b(h(S))$ will be so with respect to $\bar{d}$ and hence extend uniquely to an $f' \in C_b(\overline{h(S)})$. Conversely, given $f' \in C_b(h\overline{h(S)})$ let $\bar{f}$ be its restriction to h(S) and define $f \in C_b(S)$ by $f = \bar{f} \circ h$. Then $\bar{f}$ is uniformly continuous with respect to $\bar{d}$ and hence f is with respect to d'. Let $C_u(S)$ = the space of $f : S \to R$ which are uniformly continuous with respect to d', and endow $C_u(S)$, $C_b(\overline{h(S)})$ with the supremum norm. Then $f \longleftrightarrow f'$ as above is an isometric isomorphism between the two. Since $\overline{h(S)}$ is compact, $C_b(\overline{h(S)})$ is separable and hence $C_u(S)$ is separable. Let $\{f_n\}$ be countable dense in $C_u(S)$. If for $\{P_n\}$, P in $\mathbb{P}(S)$ we have $\int f_i\, dP_n \to \int f_i\, dP$ for all i, then we have $\int fdP_n \to \int fdP$ for all $f \in C_u(S)$ by a simple approximation argument, implying that $P_n \to P$ in $\mathbb{P}(S)$. Thus $\{f_n\}$ constitutes a convergence determining class. QED

Remarks. (1) Since the statement $\int f dP_n \to \int f dP$ does not alter if f is scaled by a nonzero scalar, it suffices to consider $\{f_n\}$ dense in the unit ball of $C_u(S)$.

(2) Yet another local base for the Prohorov topology at $P \in \mathbb{P}(S)$ is given by sets of the type (i) above with $\{f_i\}$ therein being from a prescribed countable convergence determining class and $\{\varepsilon_i\}$ rational. Considering such local bases at P in a countable dense subset of $\mathbb{P}(S)$, it follows that $\mathbb{P}(S)$ is second countable and hence metrizable.

A.3 Skorohod's Theorem

Let X_n, $n = 1,2,...,\infty$, be a collection of S-valued random variables not necessarily defined on the same probability space, and P_n, $n = 1,2...,\infty$ their laws. We say that $X_n \to X_\infty$ in law if $P_n \to P_\infty$ in $\mathbb{P}(S)$. This section explores the relation between this convergence concept and other familiar convergence concepts.

Theorem A.3.1. Suppose X_n, $n = 1,2,...,\infty$ above are defined on a common probability space and $X_n \to X_\infty$ in probability. Then $P_n \to P_\infty$ in $\mathbb{P}(S)$.

Proof. Let $f : S \to R$ be bounded and continuous uniformly with respect to d. Let $\varepsilon > 0$ and pick $\delta > 0$ such that $d(x, y) < \delta$ implies $|f(x) - f(y)| < \varepsilon$ for all $x, y \in S$. Then

$$|\int f dP_n - \int f dP_\infty| = |E[f(X_n) - f(X_\infty)]|$$

$$\leq E[|f(X_n) - f(X_\infty)| I\{d(X_n, X_\infty) < \delta\}] + E[|f(X_n) - f(X_\infty)| I\{d(X_n, X_\infty) \geq \delta\}]$$

$$\leq \varepsilon + 2 \sup_x |f(x)| P(d(X_n, X_\infty) \geq \delta)$$

$$\to \varepsilon \text{ as } n \to \infty.$$

Since $\varepsilon > 0$ was arbitrary, $\int f dP_n \to \int f dP$. The claim now follows from Corollary A.2.1 (v).

QED

A converse to this does not in general make sense as $\{X_n\}$ need not even be defined on the same probability space. A sort of partial converse is given by the following theorem due to Skorohod.

Theorem A.3.2. Let $P_n \to P_\infty$ in $\mathbb{P}(S)$. Then there exists a probability space $(\Omega, \mathcal{F}, P)$ on which there are S-valued random variables X_n, $n = 1,2,...,\infty$, such that the law of X_n is P_n for each n and $X_n \to X_\infty$ a.s.

Proof. Take $\Omega = [0,1]$, $\mathcal{F}$ its Borel σ-field and P the Lebesgue measure. To every finite collection $\{i_1, i_2,\ldots,i_k\}$, $k \geq 1$, of natural numbers, associate a Borel set $S_{(i_1,i_2,\ldots,i_k)}$ in S as follows.

(i) If $\{i_1, i_2,\ldots,i_k) \neq \{j_1, j_2,\ldots,j_k\}$, then $S_{(i_1,\ldots,i_k)} \cap S_{(j_1,\ldots,j_k)} = 0$.

(ii) $\bigcup_{j=1}^{\infty} S_j = S$, $\bigcup_{j=1}^{\infty} S_{(i_1,i_2,\ldots,i_k,j)} = S_{(i_1,i_2,\ldots,i_k)}$.

(iii) The diameter of $S_{(i_1,\ldots,i_k)} < 2^{-k}$.

(iv) $P_n(\partial S_{(i_1,\ldots,i_k)}) = 0$ for $n = 1,2,..,\infty$.

Thus for each k, $\{S_{(i_1,i_2,\ldots,i_k)}\}$ is a disjoint cover of S which is a refinement of the corresponding cover for $k' < k$. One way to obtain these sets is as follows. For each k, let B_{mk}, $m = 1,2,\ldots$, be open balls of radius not exceeding $2^{-(k+1)}$ covering S and satisfying $P_n(\partial B_{mk}) = 0$ for all n, k, m. Let $D_{1k} = B_{1k}$, $D_{nk} = B_{nk} \setminus (\bigcup_{m=1}^{n-1} B_{mk})$ and $S_{(i_1,\ldots,i_k)} = \bigcap_{j=1}^{k} D_{i_j j}$. Then (i) - (iv) hold. For each k, order $(i_1, i_2,\ldots,i_k)$ lexicographically. Define intervals $\Delta^n_{(i_1,\ldots,i_k)}$ of the form $[a, b)$ in $[0,1)$ such that

(a) the length of $\Delta^n_{(i_1,\ldots,i_k)}$ is $P_n(S_{(i_1,\ldots,i_k)})$ for $n = 1,2,\ldots,\infty$,

(b) $\bigcup_{(i_1,\ldots,i_k)} \Delta^n_{(i_1,\ldots,i_k)} = [0,1)$ for $n = 1,2,\ldots,\infty$ and

(c) whenever $(i_1, i_2,\ldots,i_k) < (j_1,\ldots,j_k)$, the interval $\Delta^n_{(i_1,\ldots,i_k)}$ lies on the left of $\Delta^n_{(j_1,\ldots,i_k)}$ for $n = 1,2,\ldots,\infty$. These intervals are uniquely determined.

For each $S_{(i_1,i_2,\ldots,i_k)}$ such that $P_n(S_{(i_1,i_2,\ldots,i_k)}) > 0$ for some n, the interior is nonempty by (iv) above, and thus we may pick a point $x_{(i_1,\ldots,i_k)}$ in the interior. For $\omega \in \Omega$, define

$$X^k_n(\omega) = \sum_{(i_1,\ldots,i_k)} x_{(i_1,\ldots,i_k)} I_{\Delta^n_{(i_1,\ldots,i_k)}}(\omega), \quad k \geq 1, \ n = 1,2,\ldots,\infty.$$

Then $d(X^k_n(\omega), X^{k+p}_n(\omega)) \leq 2^{-k}$, $p \geq 1$, $n = 1,,\ldots,\infty$, making $\{X^k_n(\omega)\}$ a Cauchy sequence for each fixed n, ω. Thus $X_n(\omega) = \lim_{k\to\infty} X^k_n(\omega)$ exists for each n, ω. Letting $|I|$ denote the length of an interval I, we have $P_n(S_{(i_1,\ldots,i_k)}) = |\Delta^n_{(i_1,\ldots,i_k)}| \to |\Delta^\infty_{(i_1,\ldots,i_k)}| = P_\infty(S_{(i_1,\ldots,i_k)})$. Hence for each w in the interior of $\Delta^\infty_{(i_1,\ldots,i_k)}$, there exists an integer $n_k(w) \geq 1$ such that w is in the interior of $\Delta^n_{(i_1,\ldots,i_k)}$ for each $n \geq n_k(w)$. Then $X^k_n(w) = X^k_\infty(w)$ for each $n \geq n_k(w)$, implying

$$\begin{aligned} d(X_n(w), X_\infty(w)) &\leq d(X_n(w), X^k_n(w)) + d(X^k_n(w), X^k_\infty(w)) \\ &\quad + d(X^k_\infty(w), X_\infty(w)) \\ &\leq 2^{-(k-1)}, \quad n \geq n_k(w). \end{aligned}$$

Set

$$\Omega_0 = \bigcap_{k=1}^{\infty} \bigcup_{(i_1,\ldots,i_k)} \text{interior } (\Delta^\infty_{(i_1,\ldots,i_k)}).$$

Then $P(\Omega_0) = 1$ and $X_n(w) \to X_\infty(w)$ for $w \in \Omega_0$. Note that for each integer $p \geq 1$,

$$\begin{aligned} P(X^{k+p}_n(w) \in \bar{S}_{(i_1,\ldots,i_k)}) &= P(X^{k+p}_n(w) \in \text{interior } (S_{(i_1,\ldots,i_k)})) \\ &= P_n(S_{(i_1,\ldots,i_k)}). \end{aligned}$$

Furthermore, every open set in S is expressible as a disjoint countable union of $S_{(i_1,\ldots,i_k)}$'s. Thus Fatou's lemma yields $\liminf_{k\to\infty} P(X^k_n \in O) \geq P_n(O)$, $O \subset S$ open. Thus the law of X^k_n converges to P_n in $\mathbb{P}(S)$ as $k \to \infty$, $n = 1,2,\ldots,\infty$. Since it clearly converges to the law of X_n as well, the law of X_n must be P_n for $n = 1,\ldots,\infty$. This completes the proof. QED

Let S′ be another Polish space.

Corollary A.3.1. Let $f : S \to S'$ be continuous and X_n, $n = 1,2,\ldots,\infty$ be S-valued random variables such that $X_n \to X_\infty$ in law. Then $f(X_n) \to f(X_\infty)$ in law as S′-valued random variables.

This is immediate from the above theorem.

A.4 Compactness in $\mathbb{P}(S)$

The following theorem of Prohorov gives a useful characterization of relative compactness in

$\mathbb{P}(S)$.

Theorem A.4.1. A subset $\Lambda \subset \mathbb{P}(S)$ is relatively compact if and only if it is tight.

Proof. Suppose S is compact. By the Riesz theorem, $\mathbb{P}(S) = \{\mu \in C^*(S) \mid \mu(f) \geq 0$ for $f \geq 0$ and $\mu(\bar{1}) = 1\}$ where $\bar{1}$ is the constant function identically equal to one. Thus $\mathbb{P}(S)$ is a weak*-closed subset of the unit ball of $C^*(S)$ and is therefore weak*-compact. Since the weak* topology of $C^*(S)$ relativized to $\mathbb{P}(S)$ coincides with the Prohorov topology, $\mathbb{P}(S)$ is compact. Suppose S is not compact. Recall the map h of section A.1. Let $\{\mu_n\}$ be a sequence in a tight set $\Lambda \subset \mathbb{P}(S)$. For each n, define $\bar{\mu}_n \in \mathbb{P}(\overline{h(S)})$ by $\bar{\mu}_n(A) = \mu_n(h^{-1}(A \cap h(S)))$ for A Borel in $(\overline{h(S)})$. By the foregoing, $\{\bar{\mu}_n\}$ is relatively compact in $\mathbb{P}(\overline{h(S)})$. Hence given any subsequence of $\{n\}$, denoted by $\{n\}$ again, there exists a further subsequence $\{n(k)\}$ such that $\bar{\mu}_{n(k)} \to \bar{\mu}$ in $\mathbb{P}(\overline{h(S)})$. For each $m = 1,2,\ldots$, let $K_m \subset S$ be a compact set such that $\mu_n(K_m) > 1-1/m$ for all n. Then $h(K_m)$ is compact for each m and

$$\bar{\mu}(h(k_m)) \geq \limsup_{k \to \infty} \bar{\mu}_{n(k)}(h(K_m))$$

$$= \limsup_{k \to \infty} \mu_{n(k)}(K_m) \geq 1 - 1/m.$$

Thus $\bar{\mu}(\bigcup_m h(K_m)) = 1 = \bar{\mu}(h(S))$ and $\bar{\mu}$ restricts to a $\mu' \in \mathbb{P}(h(S))$ on $h(S)$. Letting $\mu \in \mathbb{P}(S)$ denote the image of μ' under h^{-1}, we have

$$\limsup_{k \to \infty} \mu_{n(k)}(F) = \limsup_{k \to \infty} \bar{\mu}_{n(k)}(h(F)) \leq \bar{\mu}(h(F)) = \mu(F)$$

for all closed $F \subset S$. Thus $\mu_{n(k)} \to \mu$ in $\mathbb{P}(S)$, proving that Λ is relatively compact whenever it is tight. Conversely, let Λ be relatively compact. Let $\{x_i\} \subset S$ be a countable dense set and for $k, n \geq 1$, define $G_k^n = \bigcup_{j=1}^{n} B(x_j, 1/k)$, $B(x,r)$ being the open ball of radius r centred at x. Then the map $\mu \to \mu(G_k^n)$, $\mu \in P(S)$, is lower semicontinuous by (iii) of Corollary A.2.1, and for each fixed k, increases to $\bar{1}$ as $n \to \infty$. Since $\bar{\Lambda}$ is compact, Dini's theorem implies that for each $\varepsilon > 0$ and $k \geq 1$, there is an $n(k) \geq 1$ such that

$$\inf_{\mu \in \bar{\Lambda}} \mu(G_k^{n(k)}) \geq 1 - \varepsilon\, 2^{-k}.$$

Then for $K = \bigcap_{k=1}^{\infty} \overline{G_k^{n(k)}}$, we have

$$\inf_{\mu \in \bar{\Lambda}} \mu(K) \geq 1 - \varepsilon.$$

K is closed. Also, for any $k \geq 1$, $K \subset \bigcup_{i=1}^{n(k)} B(x_i, 1/k)$. Thus it is totally bounded, and therefore it is compact. This proves that Λ is tight. QED

The following theorem gives a useful criterion for relative compactness.

Theorem A.4.2. Let P_α, $\alpha \in I$, be a family in $\mathbb{P}(S)$ which is absolutely continuous with respect to a $P \in \mathbb{P}(S)$ with Radon-Nikodym derivatives Λ_α, $\alpha \in I$, resp. Then $\{P_\alpha, \alpha \in I\}$ is relatively compact in $\mathbb{P}(S)$, whenever $\{\Lambda_\alpha, \alpha \in I\}$ are uniformly integrable with respect to P.

Proof. This follows from the Dunford-Pettis criterion for weak compactness/sequential weak compactness in $L_1(P)$. QED

A somewhat related result is the following.

Theorem A.4.3. (Scheffe) Let P_n, $n = 1,2,\ldots,\infty$ be a probability measure on a measurable space $(\Omega, \mathcal{F})$ and λ a non-negative (but not necessarily finite) measure on $(\Omega, \mathcal{F})$ such that P_n is absolutely continuous with respect to λ for each $n = 1,2,\ldots,\infty$. If

$$\frac{dP_n}{d\lambda} \to \frac{dP_\infty}{d\lambda} \quad \lambda\text{ - a.e.},$$

then $P_n \to P_\infty$ in total variation.

Proof. This is an immediate consequence of the fact that a.e. convergence of non-negative integrable functions implies their L_1-convergence to the same limit if and only if their L_1 norms converge to the norm of the limit. QED

A.5 Metrics on $\mathbb{P}(S)$

We have already seen that $\mathbb{P}(S)$ is metrizable. Let $\{f_n\} \subset C_b(S)$ be a countable convergence determining class. It is easily seen that the following are metrics on $\mathbb{P}(S)$ consistent with the Prohorov topology.

$$d_1(\mu, \nu) = \sum_{n=1}^{\infty} 2^{-n} |\int f_n \, d\mu - \int f_n \, d\nu| \, \|f_n\|^{-1},$$

$$d_2(\mu, \nu) = \sum_{n=1}^{\infty} 2^{-n} |\int f_n \, d\mu - \int f_n \, d\nu| \wedge 1,$$

$$d_3(\mu, \nu) = \sum_{n=1}^{\infty} 2^{-n} |\int f_n \, d\mu - \int f_n \, d\nu| \, / \, (1 + |\int f_n \, d\mu - \int f_n \, d\nu|).$$

A sequence which is Cauchy under any one of these will also be Cauchy under the rest. So it suffices to test any one of them for completeness.

Theorem A.5.1. The above metrics are complete if and only if S is compact.

Proof. Let $g_n = f_n \|f_n\|^{-1}$, $n \geq 1$. The map f that maps $\mu \in \mathbb{P}(S)$ to $[\int g_1 \, d\mu, \int g_2 \, d\mu, ...]$ will map $\mathbb{P}(S)$ onto a subset of $[-1, 1]^{\infty}$. Argue as in Theorem A.1.1 to conclude that $f(\mathbb{P}(S))$ is G_{δ} in $[-1, 1]^{\infty}$ and f is a homeomorphism onto its range. Metrize $[-1,1]^{\infty}$ by the metric

$$\rho((x_1, x_2, ...), (y_1, y_2, ...)) = \sum_{n=1}^{\infty} 2^{-n} |x_n - y_n|.$$

Then f is an isometry between $(\mathbb{P}(S), d_1)$ and $(f(\mathbb{P}(S)), \rho)$. Thus d_1 is a complete metric on $\mathbb{P}(S)$ if and only if ρ is on $f(\mathbb{P}(S))$. The latter holds if and only if $f(\mathbb{P}(S))$ is closed and hence compact in $[-1, 1]^{\infty}$ which in turn is true if and only if $\mathbb{P}(S)$ is compact. By Prohorov's theorem, $\mathbb{P}(S)$ is compact when S is. Conversely, if S is not compact, it contains a sequence $\{s_n\}$ which does not have any limit point. The Dirac measures at $\{s_n\}$ are then seen to have no limit point in $\mathbb{P}(S)$ either, proving that $\mathbb{P}(S)$ is not compact. Thus $\mathbb{P}(S)$ is compact if and only if S is, completing the proof. QED

The metrics d_1, d_2, d_3 are convenient to work with, but suffer from lack of completeness for noncompact S. An alternative metric, complete regardless of whether S is compact or not, is described below. In particular, it shows that $\mathbb{P}(S)$ is Polish.

For any $\varepsilon > 0$ and any Borel set $A \subset S$, let $A^{\varepsilon} = \{x \in S \mid d(x, A) < \varepsilon\}$. For $\mu, \nu \in \mathbb{P}(S)$, define

$$q(\mu, \nu) = \inf\{\varepsilon > 0 \mid \mu(A) \leq \nu(A^{\varepsilon}) + \varepsilon, \; \nu(A) \leq \mu(A^{\varepsilon}) + \varepsilon \text{ for all Borel subsets } A \text{ of } S\}.$$

Theorem A.5.2. q defines a metric on $\mathbb{P}(S)$ consistent with the Prohorov topology.

Proof. Clearly, $q(\mu, \nu) = q(\nu, \mu) \geq 0 = q(\mu, \mu)$. Also, for closed $A \subset S$, $\mu(A^\varepsilon) \to \mu(A)$ as $\varepsilon \to 0$. Thus $q(\mu, \nu) = 0$ implies $\mu(A) = \nu(A)$ for all closed $A \subset S$, implying $\mu = \nu$. Finally, let $\mu_1, \mu_2, \mu_3 \subset \mathbb{P}(S)$. Then for any $\varepsilon > q(\mu_1, \mu_2)$ and $\delta > q(\mu_2, \mu_3)$, we have, for all Borel sets $A \subset S$,

$$\mu_1(A) < \mu_2(A^\varepsilon) + \varepsilon, \ \ \mu_2(A^\varepsilon) < \mu_3(A^{\varepsilon+\delta}) + \delta,$$

implying $\mu_1(A) < \mu_3(A^{\varepsilon+\delta}) + \varepsilon + \delta$. Similarly, $\mu_3(A) < \mu_1(A^{\varepsilon+\delta}) + \varepsilon + \delta$. Thus $q(\mu_1, \mu_3) \leq \varepsilon + \delta$. Given the choice of ε, δ, we have

$$q(\mu_1, \mu_3) \leq q(\mu_1, \mu_2) + q(\mu_2, \mu_3).$$

Thus q is a metric. Now recall the local bases (i)–(v) for the Prohorov topology described in section A.2. Let $P \in \mathbb{P}(S)$, $F \subset S$ closed and $\varepsilon > 0$. Pick $\delta \in (0, \varepsilon)$ such that $P(\partial F^\delta) = 0$ and $P(F^\delta) < P(F) + \varepsilon$. If $q(P, Q) < \delta$, $Q(F) < P(F^\delta) + \delta < P(F) + 2\varepsilon$. Thus each set of type (ii) contains an open q-ball. Next pick $\delta \in (0, \varepsilon/3)$. Cover S by open balls $\{S_i\}$ such that $P(\partial S_i) = 0$ and the diameter of $S_i < \delta$ for all i. Let $A_1 = S_1$ and $A_n = S_n \backslash (\bigcup_{m=1}^{n-1} S_m)$, $n \geq 2$. Take $k \geq 1$ such that $P(\bigcup_{i=1}^{k} A_i) > 1 - \delta$. Clearly, $P(\partial(\bigcup_{i=1}^{k} A_i)) = 0$. Take a neighbourhood N of P of type (iv) given by $N = \{Q \in \mathbb{P}(S) \mid |P(A) - Q(A)| < \delta$ for all A which can be written as a union of sets from $\{A_1, A_2, \dots, A_k\}\}$. Then $Q(\bigcup_{i=1}^{k} A_i) > 1 - 2\delta$ for $Q \in N$. For any Borel set $B \subset S$, let A denote the union of sets in $\{A_1, \dots, A_k\}$ for which $B \cap A_i \neq 0$. Then $|Q(A) - P(A)| < \delta$ for $Q \in N$. But $B \subset A \cup (\bigcup_{i=1}^{k} A_i)^c$. Since the diameter of A_i is strictly less than δ for all i, $A \subset B^\delta$ and thus

$$P(B) \leq P(A) + P((\bigcup_{i=1}^{k} A_i)^c) \leq P(A) + \delta$$

$$\leq Q(A) + 2\delta \leq Q(B^\delta) + 2\delta.$$

$$Q(B) \leq Q(A) + Q((\bigcup_{i=1}^{k} A_i)^c) \leq Q(A) + 2\delta$$

$$\leq P(A) + 3\delta \leq P(B^\delta) + 3\delta.$$

Hence $q(P,Q) \le 3\delta < \varepsilon$. Thus every open q-ball contains a set of type (iv). This completes the proof. QED

Theorem A.5.3. q is a complete metric on $\mathbb{P}(S)$.

Proof. Let $\{P_n\} \subset \mathbb{P}(S)$ be a Cauchy sequence with respect to q. Let $\varepsilon > 0$ and $k \ge 1$. Pick $n(k) \ge 1$ such that for $n \ge n(k)$, $q(P_n, P_{n(k)}) < \varepsilon\, 2^{-k}$. Let $B_1^k, \dots, B_{m(k)}^k$ be finitely many open balls of diameter $\varepsilon\, 2^{-k}$ satisfying

$$P_{n(k)}\Big(\bigcup_{i=1}^{m(k)} B_i^k\Big) > 1 - \varepsilon\, 2^{-k}.$$

Let $A_1^k, \dots, A_{m(k)}^k$, be open balls with the same centres as the corresponding B_i^k's and with diameter $\varepsilon\, 2^{-k+1}$. Since

$$\Big(\bigcup_{i=1}^{m(k)} B_i^k\Big)^{\varepsilon 2^{-k}} \subset \bigcup_{i=1}^{m(k)} A_i^k$$

and $q(P_n, P_{n(k)}) < \varepsilon\, 2^{-k}$ for $n \ge n(k)$, we have

$$P_n\Big(\bigcup_{i=1}^{m(k)} A_i^k\Big) > 1 - \varepsilon\, 2^{-k+1}, \quad n \ge n(k).$$

By including finitely many additional open balls of diameter $\varepsilon\, 2^{-k+1}$ if necessary, say $A_{m(k)+1}^k, \dots, A_{j(k)}^k$, we have

$$P_n\Big(\bigcup_{i=1}^{j(k)} A_i^k\Big) > 1 - \varepsilon\, 2^{-k+1} \text{ for all } n.$$

Let $K = \bigcap_{k \ge 1} \bigcup_{i \le j(k)} A_i^k$. Then $\bar{K}$ is totally bounded and closed, and hence compact. Also, $P_n(\bar{K}) > 1 - 2\varepsilon$ for all n. Thus $\{P_n\}$ is tight, and hence relatively compact in $\mathbb{P}(S)$. Since it is Cauchy as well, it must converge. QED

Finally, here's yet another metric for $\mathbb{P}(S)$ which provides some additional intuition concerning the convergence in $\mathbb{P}(S)$. Define

$$\bar{q}(\mu, \nu) = \inf E\,[d(X, Y)\,], \quad \mu, \nu \in \mathbb{P}(S), \tag{A.1}$$

where the infimum is over all pairs (X, Y) defined on some probability space (not necessarily the same for two such pairs) with the law of X being μ and that of Y being ν. Before studying $\bar{q}$ further, we insert a technical lemma.

Lemma A.5.1. Let $\mu_1, \mu_2 \in \mathbb{P}(S \times S)$ be such that the image ν of μ_1 under the projection onto the second factor space coincides with the image of μ_2 under the projection onto the first factor space (i.e., the measures $\mu_1(S, dy)$ and $\mu_2(dy, S)$ constitute the same element ν of $\mathbb{P}(S)$). Then there exist S-valued random variables X, Y, Z defined on some probability space $(\Omega, \mathcal{F}, P)$ such that the law of (X, Y) is μ_1 and the law of (Y, Z) is μ_2.

Proof. Anticipating the results of the next section, we disintegrate μ_1, μ_2 as

$$\mu_1(dx, dy) = \nu(dy)v_1(y, dx), \mu_2(dy, dz) = \nu(dy)v_2(y, dz)$$

where v_1, v_2 are the appropriate regular conditional laws defined ν-a.s. Let $\Omega = S \times S \times S$ and $\mathcal{F}$ its Borel σ-field. Define a probability measure P on $(\Omega, \mathcal{F})$ by

$$P(A \times B \times C) = \int_B \int_A \int_C v_2(y, dz)\, v_1(y, dx)\, \nu(dy)$$

for A, B, C Borel in S. Letting $w = (w_1, w_2, w_3)$ denote a typical element of Ω, the random variables X, Y, Z on $(\Omega, \mathcal{F}, P)$ defined by $X(w) = w_1$, $Y(w) = w_2$, $Z(w) = w_3$ will serve the purpose. QED

Theorem A.5.4. $\bar{q}$ is a metric on $\mathbb{P}(S)$ consistent with the Prohorov topology.

Proof. Clearly, $\bar{q}(\mu, \nu) = \bar{q}(\nu, \mu) \geq 0 = \bar{q}(\mu, \mu)$. (Take $X = Y$ in (A.1).) Suppose $\bar{q}(\mu, \nu) = 0$. Then there exist probability spaces $(\Omega_n, \mathcal{F}_n, P_n)$, $n = 1,2,\ldots$, on which there are random variables X_n, Y_n respectively such that the law of X_n (resp. Y_n) is μ (resp. ν) for each n and $E[d(X_n, Y_n)] \to 0$. By mimicking the arguments of Theorem A.3.1, one can show that

$$\left| \int f d\mu - \int f d\nu \right| = \left| E[f(X_n) - f(Y_n)] \right| \to 0$$

as $n \to \infty$ for $f \in C_b(S)$ which are uniformly continuous with respect to d. Thus $\mu = \nu$. Next, let $\varepsilon > 0$ and let (X_1, Y_1), (Y_2, Z_2) be $S \times S$-valued random variables (not necessarily on the same probability space) such that the laws of X_1, Y_1, Y_2, Z_2 are $\mu_1, \mu_2, \mu_2, \mu_3$ respectively with

$$\bar{q}(\mu_1, \mu_2) + \varepsilon \geq E[d(X_1, Y_1)],$$

$$\bar{q}(\mu_2, \mu_3) + \varepsilon \geq E[d(Y_2, Z_2)].$$

By Lemma A.5.1 we can construct on some probability space random variables X, Y, Z such that the laws of (X, Y) and (X_1, Y_1) coincide and the laws of (Y, Z) and (Y_2, Z_2) coincide. Then

$$\bar{q}(\mu_1, \mu_3) \leq E[d(X, Z)] \leq E[d(X, Y)] + E[d(Y, Z)]$$

$$\leq \bar{q}(\mu_1, \mu_2) + \bar{q}(\mu_2, \mu_3) + 2\varepsilon .$$

Letting $\varepsilon \to 0$, we get the triangle inequality. Thus $\bar{q}$ is a metric on $\mathbb{P}(S)$. Let $\mu_n \to \mu_\infty$ in $\mathbb{P}(S)$ with the Prohorov topology. By Skorohod's theorem, there exist random variables X_n, $n = 1,2,\ldots,\infty$, on some probability space such that the law of X_n is μ_n for each n and $X_n \to X_\infty$ a.s. Then $E[d(X_n, X_\infty)] \to 0$, implying $\bar{q}(\mu_n, \mu_\infty) \to 0$. Conversely, let μ_n, $n = 1,2,\ldots,\infty$ be a sequence in $\mathbb{P}(S)$ such that $\bar{q}(\mu_n, \mu_\infty) \to 0$. Then there exist probability spaces $(\Omega_n, \mathcal{F}_n, P_n)$, $n \geq 1$, on which there are random variables X_n, Y_n respectively such that the law of X_n is μ_n and that of Y_n is μ_∞ for each n, and $E[d(X_n, Y_n)] \to 0$. Once again one can mimic the arguments of Theorem A.3.1 to conclude that $\mu_n \to \mu_\infty$ in the Prohorov topology of $\mathbb{P}(S)$. This completes the proof. QED

Theorem A.5.5. The metric $\bar{q}$ is complete.

Proof. Let $\{\mu_n\} \subset \mathbb{P}(S)$ be Cauchy under $\bar{q}$. Let $A \subset S$ be Borel, $n, m \geq 1$ and X, Y any S-valued random variables on a common probability space with laws μ_n, μ_m respectively. Then

$$\mu_n(A) - \mu_m(A^\delta) = E[I\{X \in A\} - I\{Y \in A^\delta\}]$$

$$\leq P(X \in A,\ Y \notin A^\delta)$$

$$\leq P(d(X, Y) \geq \delta)$$

$$\leq \delta^{-1} E[d(X, Y)].$$

Taking the infimum over all choices of (X, Y) on the right, one gets

$$\mu_n(A) - \mu_m(A^\delta) \leq \delta^{-1} \bar{q}(\mu_n, \mu_m).$$

For n,m sufficiently large, the right hand side is less than δ. A symmetric argument shows

$\mu_m(A) < \mu_n(A^\delta) + \delta$. From the definition of q, we conclude that $q(\mu_n, \mu_m) < \delta$, for n, m sufficiently large. Thus $\{\mu_n\}$ is Cauchy under q and hence $q(\mu_n, \mu) \to 0$ for some $\mu \in \mathbb{P}(S)$. In view of the preceding theorem, $\bar{q}(\mu_n, \mu) \to 0$. Thus $\bar{q}$ is complete. QED

A.6 Disintegration of Measures

This section establishes in particular the result, already used once, that a probability measure μ on $S_1 \times S_2$ (S_1 and S_2 being prescribed Polish spaces) 'disintegrates' as $\mu(dx, dy) = \bar{\mu}(dx)v(x, dy)$ with $\bar{\mu}$ = the image of μ under the projection onto S_1 and $x \to v(x,.) : S_1 \to \mathbb{P}(S_2)$ is $\bar{\mu}$ - a.s. unique. This is a special case of the following more general result:

Theorem A.6.1. Let X be an S-valued random variable on a probability space $(\Omega, \mathcal{F}, P)$ and let $\bar{\mathcal{F}}$ be a sub-σ-field of $\mathcal{F}$. Then there exists an a.s. unique $\bar{\mathcal{F}}$-measurable map $w \to v(w,.) : \Omega \to \mathbb{P}(S)$ such that for any $f \in C_b(S)$,

$$E[f(X)/\bar{\mathcal{F}}] = \int f(x)v(.,dx) \quad P\text{-a.s.}$$

Remarks. (1) To see that this indeed implies the above mentioned disintegration of μ, let $\Omega = S_1 \times S_2$, $\mathcal{F}$ the product σ-field and $P = \mu$. Define $X : \Omega \to S_2$ by $X((w_1, w_2)) = w_2$ for $(w_1, w_2) \in \Omega$ and $\bar{\mathcal{F}} = \{A \times S_2 \mid A \text{ Borel in } S_1\}$. The rest is easy.

(2) The function v above is called the regular conditional law of X given $\bar{\mathcal{F}}$ (or X given Y when $\bar{\mathcal{F}}$ is the σ-field generated by a random variable Y).

Proof. Recall d', $C_u(S)$ from Theorem A.2.2. Let $\{f_n\}$ be countable dense in $C_u(S)$ with $f_1(.) \equiv 1$ and N the set of zero probability outside which

$$E[\sum_{i=1}^{m} a_i f_i(X)/\bar{\mathcal{F}}] = \sum_{i=1}^{m} a_i E[f_i(X)/\bar{\mathcal{F}}] \tag{A.2}$$

for all choices of $m \geq 1$ and rationals $\{a_1,\dots,a_m\}$. Then for each sample point $w \notin N$, (A.2) defines a positive bounded linear functional on the linear span of $\{f_i\}$ over the field of rationals, viewed as a dense subspace of $C_u(S)$. This extends uniquely to a positive bounded linear functional on $C_u(S)$. This can be identified with a positive bounded linear functional of $C(\overline{h(S)})$ via the scheme of Theorem A.2.2. That in turn can be identified with a positive finite measure μ on $\overline{h(S)}$ Also, $\mu(\bar{1}) = E[f_1(X)/\bar{\mathcal{F}}] = 1$ a.s. We may drop the 'a.s.' here by enlarging N if necessary. Thus $\mu \in \mathbb{P}(\overline{h(S)})$. This procedure defines $\mu : \Omega\backslash N \to \mathbb{P}(\overline{h(S)})$.

The definition can be extended to $\mu : \Omega \to \mathbb{P}(\overline{h(S)})$ by setting μ equal to an arbitrary fixed element of $P(\overline{h(S)})$ on N. The measurability of this map is seen to be equivalent to that of $\int f d\mu = E[f(X)/\bar{\mathcal{F}}]$ for all $f \in C(\overline{h(S)})$. Thus μ is an $\bar{\mathcal{F}}$ measurable $\mathbb{P}(\overline{h(S)})$-valued random variable. Next, note that $E[\mu(h(S))] = P(X \in S) = 1$, implying that $\mu(h(S)) = 1$ a.s. By modifying μ on a set of zero probability if necessary, we can drop the 'a.s.' and identify μ with its restriction to $h(S)$. Let $v(w,.) \in \mathbb{P}(S)$ denote the pullback of μ under h. The claim now follows by direct verification. The a.s. uniquenes is easily verified. QED

For a more advanced treatment of the spaces of probability measures and disintegration of measures, see [Par] and [Schw] respectively.

References

[ABB] K.J. Arrow, E.W. Barankin and D. Blackwell, 'Admissible points of convex sets', in *Contributions to the Theory of Games,* Vol. II, H.W. Kuhn and A.W. Tucker (eds.), (Princeton University Press, Princeton, 1950), 87-91.

[AlSh] E. Altman and A. Shwartz, *Sensitivity of constrained Markov decision processes,* (Dept. of Electrical Eng., Technion, Haifa, Israel, Jan. 1990).

[Ar Ma] A. Arapostathis and S.I. Marcus, 'Analysis of an identification algorithm arising in the adaptive estimation of Markov chains', *Math. of Control, Signals and Systems* 1 (1990), 1-29.

[Ben] V.E. Benes, 'Existence of optimal strategies based on specified information, for a class of stochastic decision problems', *SIAM J. Control 8* (1970), 179-188.

[BhMa] F.J. Beutler and K.W. Ross, 'Optimal policies for controlled Markov chains with a constraint', *J. Math. Analysis and Appl.*, 122 (1985), 236-252.

[BeRo] R.N. Bhattacharya and M. Majumdar, 'Controlled semi-Markov model under long-run average rewards', *J. Statistical Planning and Appl.*, 22 (1989), 223-242.

[Bill] P. Billingsley, *Convergence of probability measures*, (Wiley, New York, 1968).

[Bor 1] V.S. Borkar, *Identification and adaptive control of Markov chains*, (Ph.D. Thesis, Dept. of Electrical Eng. and Computer Sciences, University of California, Berkeley, 1980).

[Bor 2] V.S. Borkar, 'Controlled Markov chains and stochastic networks', *SIAM J. Control and Opt.* 21 (1983), 652-666.

[Bor 3] V.S. Borkar, 'On minimum cost per unit time control of Markov Chains', *SIAM J. Control and Opt.* 22 (1984), 965-978.

[Bor 4] V.S. Borkar, 'Control of Markov chains with long-run average cost criterion', in *Proc. of the I.M.A. Workshop on Stochastic Differential Systems, Stochastic Control Theory and Appl.*, W. Fleming and P.L. Lions (eds.), (Springer, New York, 1987), 57-77.

[Bor 5] V.S. Borkar, 'A convex analytic approach to Markov decision processes', *Prob. Theory and Related Fields* 78 (1988), 583-602.

[Bor 6] V.S. Borkar, 'Control of Markov chains with long-run average cost criterion: the dynamic programming equations', *SIAM J. Control and Opt.* 27 (1989), 642-657.

[Bor 7] V.S. Borkar, 'Controlled Markov chains with constraints', preprint.

[Bor 8] V.S. Borkar, 'A remark on control of partially observed Markov chains', *Annals of Op. Research* (forthcoming).

[Bor 9] V.S. Borkar, 'The Kumar-Becker-Lin scheme revisited' *J. Opt. Theory and Appl.*, 66 (1990), 289-309.

[Bor 10] V.S. Borkar, 'Controlled Markov chains with constraints II', Proc. of the Workshop on Recent Advances in Stochastic Models and Their Appl., Bangalore, Jan. 1991 (forthcoming).

[BoGh] V.S. Borkar and M.K. Ghosh, 'Ergodic and adaptive control of nearest-neighbour motions' in *Maths. of Control, Signals and Systems* (forthcoming), 1991.

[BoV 1] V.S. Borkar and P. Varaiya, 'Adaptive control of Markov chains' in *Stochastic Control Theory and Stochastic Differential Systems*, M. Kohlmann and W. Vogel (eds.), Lecture Notes in Control and Info. Sciences No. 16, (Springer, Berlin-Heidelberg, 1979), 294-296.

[BoV2] V.S. Borkar and P. Varaiya, 'Adaptive control of Markov chains I : finite parameter set', *IEEE Trans. on Automatic Control* Ac-24 (1979), 953-957.

[BoV 3] V.S. Borkar and P. Varaiya, 'Identification and adaptive control of Markov chains', *SIAM J. Control and Opt*. 20 (1982), 470-489.

[BVW] C. Buyukkoc, P. Varaiya and J. Walrand, 'Extensions of the multiarmed bandit problem: the discounted case', *IEEE Trans. on Automatic Control* AC-30 (1985), 426-439.

[Ch Te] Y.S. Chow and H. Teicher, *Probability theory: independence, interchangeability, martingales,* (Springer, New York, 1978).

[DeQ] F. Delebeque and J.P. Quadrat, 'Optimal control of Markov chains admitting strong and weak interactions', *Automatica* 17 (1981), 281-296.

[DoVa] M. Donsker and S.R.S. Varadhan, 'Asymptotic evaluation of certain Markov process expectations for large time III', *Comm. in Pure and Applied Math*. XXIX (1976), 389-461.

[FHT] A. Federgruen, A. Hordijk and H.C. Tijms, 'A note on simultaneous recurrence conditions on a set of denumerable stochastic matrices', *J. Appl. Prob.* 15 (1978), 842-847.

[FoVa] J.-P. Forestier and P. Varaiya, 'Multilayer control of large Markov chains', *IEEE Trans. on Automatic Control* AC-23 (1978), 298-304.

[Gh] M.K. Ghosh, 'Markov decision processes with multiple costs', *Op. Research Letters* 9 (1990), 257-260.

[Haj] B. Hajek, 'Hitting-time and occupation-time bounds implied by drift analysis with applications', *Adv. Appl. Prob.* 14 (1982), 502-525.

[He Ma] O. Hernandez-Lerma and S.I. Marcus, 'Adaptive control of Markov processes with incomplete state information and unknown parameters', *J. Opt. Theory and Appl*. 52 (1987), 227-241.

[Hs Ma] K. Hsu and S.I. Marcus, 'Decentralized control of finite state Markov processes', *IEEE Trans. on Automatic Control* AC-27 (1982), 426-431.

[Is Ma] D.L. Isaacson and R.W. Madsen, *Markov chains: theory and applications*, (Wiley, New York, 1976).

[Kar] S. Karlin, *Mathematical methods and theory in games, programming and economics*, Vol. I, (Addison-Wesley, Reading, Mass., 1959).

[Kum] P.R. Kumar, 'Survey of results in stochastic adaptive control', *SIAM J. Control and Opt*. 23 (1985), 329-380.

[Ku Be] P.R. Kumar and A. Becker, 'A new family of optimal adaptive controllers for Markov chains', *IEEE Trans. on Automatic Control* AC-27 (1982), 137-145.

[Ku Li] P.R. Kumar and W. Lin, 'Optimal adaptive controllers for unknown Markov chains', *IEEE Trans. on Automatic Control* AC-27 (1982), 765-774.

[Ku Va] P.R. Kumar and P.P. Varaiya, *Stochastic systems: estimation, identification and adaptive control*, (Prentice-Hall, Englewood Cliffs, New Jersey, 1986).

[Lo] M. Loève, *Probability theory II*, 4th ed., (Springer, New York, 1977).

[Luen] D.G. Luenberger, *Optimization by vector space methods*, (Wiley, New York, 1969).

[MVRS] I.M. Makarov, T.M. Vinogradskaya, A.A. Rubchinsky and V.B. Sokolov, *The theory of choice and decision making*, (Mir, Moscow, 1987).

[Ma Sc] A. Makowski and A. Schwartz, 'Implementation issues for Markov decision processes', in *Proc. of the IMA Workshop on Stochastic Differential Systems, Stochastic Control Theory and Appl.*, W. Fleming and P.L. Lions (eds.), (Springer, New York 1987), 323-337.

[Mand] P. Mandl, 'Estimation and control in Markov chains', *Adv. Appl. Prob.* 6 (1974), 40-60.

[Nev] J. Neveu, *Discrete parameter martingales*, (North Holland, Amsterdam, 1975).

[Par] K.R. Parthasarathy, *Probability measures on metric spaces*, (Academic Press, New York, 1967).

[Phe] R. Phelps, *Lectures on Choquet's theorem*, (Van Nostrand, New York, 1966).

[Plat] L.K. Platzman, 'Optimal infinite-horizon undiscounted control of finite probabilistic systems', *SIAM J. Control and Opt*. 18 (1980), 362-380.

[RVW] Z. Rosberg, P. Varaiya and J. Walrand, 'Optimal control of service in tandem queues', *IEEE Trans. on Automatic Control* AC-27 (1982), 600-609.

[Ross] K.W. Ross, 'Randomized and past-dependent policies for Markov decision processes with multiple constraints', *Op. Research* 37 (1989), 474-477.

[Scha] M. Schäl, 'Estimation and control in discounted dynamic programming', *Stochastics* 20 (1987), 51-71.

[Schw] L. Schwartz, *Lectures on disintegration of measures*, Notes by S. Ramaswamy, (Tata Institute of Fundamental Research, Bombay, 1976).

[Tijm] H.C. Tijms, *Stochastic modelling and analysis: a computational approach*, (Wiley, Chichester, 1986).

[VaH] K. Van Hee, 'Bayesian control of Markov chains', *Mathematisch Centrum Tracts* No. 95, (Math. Centrum, Amsterdam, 1978).

[Whi 1] P. Whittle, *Optimization over time: dynamic programming and stochastic control*, Vol. 1, (Wiley, Chicester, 1983).

[Whi 2] P. Whittle, *Optimization over time: dynamic programming and stochastic control*, Vol. 2, (Wiley, Chicester, 1983).

UW 0813500 2